Chapter 1: The Chemistry of Life

Biochemistry, the science of the chemical processes within living organisms, is the bridge between biology and chemistry. It explores the fundamental chemical structures and reactions that underpin life, driving everything from metabolism and genetics to cell communication and the immune response. Understanding biochemistry provides insight not only into the molecular machinery of life but also into how these processes manifest in health, disease, and biotechnology.

The Role of Atoms, Molecules, and Ions in Biochemistry

Life at its core is a network of chemical reactions that occur in a world of atoms, molecules, and ions. Atoms are the building blocks of matter, composed of protons, neutrons, and electrons. The interactions between these atoms form molecules, which can range from the simplest, like oxygen (O_2), to the complex, like the DNA that encodes our genetic information.

Biochemical molecules are primarily composed of a few essential elements: carbon (C), hydrogen (H), oxygen (O), nitrogen (N), phosphorus (P), and sulfur (S). These elements are often bound together in molecules that perform specific functions necessary for life. For example, DNA is a long chain of nucleotides made from carbon, nitrogen, oxygen, and phosphorus atoms. The unique properties of carbon, with its ability to form four stable covalent bonds, make it the ideal element for building the complex molecules of life.

Ions, which are atoms or molecules with an electrical charge due to the loss or gain of electrons, also play a crucial role in biochemistry. Sodium (Na^+), potassium (K^+), calcium (Ca^{2+}), chloride (Cl^-), and others help maintain the electrochemical gradients across cell membranes, critical for processes like nerve signaling and muscle contraction.

Introduction to Chemical Bonds

The behavior of atoms and molecules is governed by chemical bonds, which are the forces that hold atoms together within molecules. There are three primary types of chemical bonds: ionic, covalent, and hydrogen bonds.

- **Ionic Bonds**: These bonds form when one atom donates an electron to another, resulting in charged ions. For example, sodium (Na) can lose an electron to form Na^+, while chlorine (Cl) can gain an electron to form Cl^-. The attraction between oppositely charged ions creates an ionic bond, such as the bond between sodium and chloride in common table salt (NaCl). Ionic bonds are important in biochemistry because they help stabilize molecular structures and contribute to the formation of ion gradients across membranes.
- **Covalent Bonds**: In covalent bonds, atoms share electrons to achieve a stable electron configuration. This type of bond is prevalent in biochemistry because it allows for the formation of stable molecules. For instance, in water (H_2O), each hydrogen atom shares an electron with the oxygen atom, forming a covalent bond. This sharing is also crucial in larger biological molecules, such as proteins and nucleic acids, where covalent bonds hold amino acids and nucleotides together.
- **Hydrogen Bonds**: Hydrogen bonds are weaker interactions between a hydrogen atom attached to a highly electronegative atom (such as oxygen or nitrogen) and another electronegative atom. While these bonds are weaker than ionic and covalent bonds, they are essential in maintaining the three-dimensional structures of biological molecules. In DNA, for instance, hydrogen bonds hold together the two strands of the double helix, ensuring the stability and accessibility of genetic information.

Water, the most abundant molecule in living organisms, is often referred to as the "solvent of life" due to its unique properties that support life processes. Water's polar nature—where one side of the molecule has a partial positive charge (due to the hydrogen atoms) and the other side has a partial negative charge (due to the oxygen atom)—allows it to dissolve many ionic and polar substances. This is essential for biochemical reactions, as most reactions occur in aqueous solutions inside cells.

Water also has a high specific heat, meaning it can absorb a large amount of heat without a significant change in temperature. This property helps organisms maintain stable internal temperatures, a condition essential for the proper functioning of enzymes and other biochemical processes. Additionally, water's ability to form hydrogen bonds contributes to its high surface tension and cohesion, facilitating processes like nutrient transport in plants and blood circulation in animals.

pH, Buffers, and Acid-Base Chemistry in Biological Systems

In biochemistry, pH is a measure of the acidity or alkalinity of a solution, based on the concentration of hydrogen ions (H^+). Most biochemical reactions are highly sensitive to changes in pH, as the charge and structure of molecules can change with pH, altering their function. For example, enzymes, which are catalysts that speed up biochemical reactions, have an optimal pH at which they function best. Changes in pH outside of this range can denature enzymes, rendering them inactive.

To maintain the proper pH in cells and organs, organisms rely on **buffers**, which are systems of molecules that can absorb or release hydrogen ions to resist changes in pH. The most important buffering system in human blood is the **bicarbonate buffer**. In this system, carbonic acid (H_2CO_3) dissociates into bicarbonate (HCO_3^-) and hydrogen ions (H^+). If the pH drops (indicating an increase in H^+ concentration), bicarbonate can absorb excess hydrogen ions, raising the pH back to normal. Conversely, if the pH rises, carbonic acid can donate H^+ to restore balance.

Buffer systems are critical in maintaining the stability of biological environments, ensuring that metabolic processes can occur without disruption. This is especially important in cellular processes like protein folding, enzyme activity, and the transmission of nerve impulses.

The Importance of Biochemistry in Everyday Life

The principles of biochemistry are deeply interwoven with our everyday experiences, even though we may not always recognize them. When we eat, our body breaks down food into molecules like glucose and fatty acids, which are used to produce energy in our cells. When we exercise, biochemical reactions in our muscles fuel our movements. In our immune system, biochemical processes allow us to fight infections, and our genetic code—stored in DNA—encodes the instructions for building and maintaining our bodies.

On a broader scale, biochemistry has profound implications for medicine and technology. Advances in biochemistry have led to the development of vaccines, therapies for genetic disorders, and cutting-edge biotechnology that can produce pharmaceuticals, genetically modified crops, and biofuels. Understanding biochemistry is not just for scientists in laboratories; it is essential knowledge for anyone seeking to understand the biological world and improve the human condition.

Conclusion

In this chapter, we have introduced some of the fundamental concepts of biochemistry: atoms, molecules, ions, and chemical bonds. We also explored the properties of water and the role of pH and buffers in maintaining life's delicate balance. These concepts form the foundation of biochemistry, and throughout the book, we will build upon this knowledge to explore the complex and fascinating molecular processes that sustain life. Whether you are a student, a professional, or just a curious reader, this book aims to provide you with a comprehensive understanding of biochemistry, from its basic principles to its cutting-edge applications in medicine, technology, and beyond.

Chapter 2: Building Blocks of Life

Biochemistry, at its core, is the study of the molecular mechanisms that underlie the structure and function of biological systems. To truly understand biochemistry, it is essential to first explore the fundamental building blocks that make up all living organisms. These building blocks — amino acids, nucleic acids, carbohydrates, and lipids — are the molecules from which life is constructed. In this chapter, we will discuss each of these key components, their structures, properties, and their roles in the functioning of the cell.

Amino Acids: The Building Blocks of Proteins

Amino acids are organic compounds that serve as the building blocks for proteins. Proteins, in turn, perform a vast array of essential biological functions, from catalyzing chemical reactions (as enzymes) to serving as structural components in tissues and organs. There are 20 standard amino acids that are encoded by the genetic code, and their unique properties are determined by their side chains, also known as R-groups.

Structure of Amino Acids:

An amino acid consists of a central carbon (called the α-carbon) attached to an amino group (-NH₂), a carboxyl group (-COOH), a hydrogen atom, and a variable side chain (R-group). The side chain is what differentiates each amino acid from the others and determines its properties, such as hydrophobicity, charge, and polarity. For instance, the side chain of glycine is a single hydrogen atom, making it the smallest amino acid, while the side chain of tryptophan contains an aromatic ring structure, making it bulky and hydrophobic.

Amino acids can be classified into several categories based on their side chains:

- **Nonpolar, hydrophobic:** These amino acids have side chains that do not interact favorably with water. Examples include alanine, valine, and leucine.
- **Polar, hydrophilic:** These amino acids have side chains that can form hydrogen bonds with water. Examples include serine, threonine, and asparagine.
- **Acidic (negatively charged):** These amino acids have side chains that can donate protons (H^+) and are negatively charged at physiological pH. Examples include aspartic acid and glutamic acid.
- **Basic (positively charged):** These amino acids have side chains that can accept protons and are positively charged at physiological pH. Examples include lysine, arginine, and histidine.

Peptides and Proteins:

When two or more amino acids join together through peptide bonds (a type of covalent bond between the carboxyl group of one amino acid and the amino group of another), they form peptides. A peptide is a short chain of amino acids, while a protein is a longer, more complex peptide that often consists of multiple polypeptide chains folded into a specific three-dimensional structure.

Proteins are classified by their structure:

1. **Primary Structure:** This refers to the linear sequence of amino acids in a protein. The sequence of amino acids is determined by the DNA code, which is transcribed into mRNA and then translated into a polypeptide chain.
2. **Secondary Structure:** The polypeptide chain folds into regular structures, such as alpha-helices and beta-pleated sheets, stabilized by hydrogen bonds.
3. **Tertiary Structure:** This is the overall three-dimensional shape of a protein, formed by interactions between the side chains of amino acids (such as hydrophobic interactions, hydrogen bonding, and disulfide bridges).
4. **Quaternary Structure:** Many proteins are composed of multiple polypeptide subunits that come together to form a functional protein. Hemoglobin, for example, is a protein with quaternary structure, composed of four subunits.

Functions of Proteins:

Proteins play a diverse range of roles in living organisms. Some are enzymes that catalyze biochemical reactions, others serve as structural components in cells (like collagen and keratin), while others function as antibodies in the immune system or transport molecules across membranes (such as hemoglobin, which carries oxygen in red blood cells).

Nucleic Acids: DNA and RNA

Nucleic acids are the molecules that store and transmit genetic information in all living organisms. There are two primary types of nucleic acids: **deoxyribonucleic acid (DNA)** and **ribonucleic acid (RNA)**. Both are long polymers made up of repeating units called nucleotides, each consisting of a nitrogenous base, a sugar molecule, and a phosphate group.

DNA:

DNA is the genetic material in almost all living organisms. It is composed of two long strands of nucleotides that coil around each other to form a double helix. The four nitrogenous bases in DNA are:

- **Adenine (A)**
- **Thymine (T)**
- **Cytosine (C)**
- **Guanine (G)**

The strands are held together by hydrogen bonds between complementary base pairs: adenine pairs with thymine, and cytosine pairs with guanine. This complementary base-pairing mechanism allows for the accurate replication of DNA during cell division. DNA carries the genetic code that dictates the synthesis of proteins.

RNA:

RNA is similar to DNA but differs in several key ways. It is typically single-stranded and contains the base uracil (U) instead of thymine (T). RNA plays a crucial role in protein synthesis. The three major types of RNA are:

- **Messenger RNA (mRNA):** This type carries genetic information from DNA in the nucleus to the ribosome, where proteins are synthesized.
- **Transfer RNA (tRNA):** This molecule helps decode mRNA into a sequence of amino acids during protein synthesis.
- **Ribosomal RNA (rRNA):** This is a major component of ribosomes, the cellular structures responsible for synthesizing proteins.

Roles in Heredity and Protein Synthesis:

DNA contains the instructions for building proteins, and this genetic information is passed down from parent to offspring. The process of protein synthesis begins with the transcription of DNA into mRNA. The mRNA is then translated into a specific sequence of amino acids, which folds into a functional protein. This central dogma of molecular biology — DNA → RNA → Protein — forms the basis of genetic expression in all living organisms.

Carbohydrates: Sugars and Energy Storage

Carbohydrates are organic compounds made up of carbon, hydrogen, and oxygen. They are the body's primary source of energy and are involved in numerous metabolic processes. Carbohydrates are classified into several types based on their structure and function:

- **Monosaccharides:** These are simple sugars, such as glucose, fructose, and galactose. Glucose is the most important monosaccharide in biochemistry because it is the primary energy source for cells.
- **Disaccharides:** These are formed by the bonding of two monosaccharides. For example, sucrose (table sugar) is composed of glucose and fructose, while lactose (milk sugar) consists of glucose and galactose.
- **Polysaccharides:** These are large, complex carbohydrates made up of long chains of monosaccharides. Examples include:

- **Starch:** A storage form of glucose in plants.
- **Glycogen:** A storage form of glucose in animals, primarily in the liver and muscles.
- **Cellulose:** A structural polysaccharide found in plant cell walls.

Carbohydrates provide a readily available source of energy. When glucose is metabolized through glycolysis, it is broken down to produce ATP, the energy currency of the cell.

Lipids: Fats, Phospholipids, and Membranes

Lipids are a diverse group of hydrophobic molecules that include fats, oils, phospholipids, and steroids. Lipids play important roles in energy storage, membrane structure, and signaling.

- **Fats and Oils:** These are triglycerides, composed of one glycerol molecule bonded to three fatty acids. Fats are used for long-term energy storage, and their breakdown produces more energy per gram than carbohydrates or proteins. Unsaturated fats, found in plants and fish, contain double bonds in their fatty acid chains, while saturated fats, found in animal products, contain only single bonds.
- **Phospholipids:** These are similar to triglycerides but have only two fatty acid chains and a phosphate group attached to the glycerol backbone. Phospholipids are a key component of cell membranes, where they form bilayers that provide structural integrity and selectively allow molecules to pass in and out of the cell.
- **Steroids:** These are lipids with a structure consisting of four fused carbon rings. Cholesterol is the most well-known steroid, and it serves as a precursor for the synthesis of hormones like estrogen, testosterone, and cortisol.

Membranes:

Lipids, particularly phospholipids, form the fundamental structure of biological membranes, including the plasma membrane surrounding every cell. The hydrophobic tails of phospholipids face inward, while the hydrophilic heads face outward, creating a semi-permeable barrier that controls the movement of substances in and out of the cell.

Conclusion

The building blocks of life — amino acids, nucleic acids, carbohydrates, and lipids — are integral to the structure and function of all living organisms. Together, these molecules enable the biochemical processes that sustain life, from protein synthesis and energy storage to genetic inheritance and cell communication. By understanding the properties and functions of these fundamental components, we lay the groundwork for exploring more complex biochemical pathways in the subsequent chapters. In the next chapter, we will delve into the functioning of enzymes, the biological catalysts that facilitate nearly all biochemical reactions.

Chapter 3: Enzyme Function and Kinetics

Enzymes are fundamental to all biochemical processes, catalyzing virtually every reaction that occurs within living organisms. Their ability to accelerate chemical reactions is essential for life itself, from breaking down nutrients to synthesizing complex molecules. In this chapter, we will explore the structure and function of enzymes, their interactions with substrates, and the principles that govern enzyme kinetics.

The Structure and Function of Enzymes

Enzymes are proteins that act as biological catalysts, speeding up chemical reactions without being consumed in the process. They are highly specific in terms of the reactions they catalyze, and this specificity is determined by their three-dimensional structure. The structure of an enzyme consists of one or more polypeptide chains that fold into a precise three-dimensional shape. This shape is crucial because it allows the enzyme to interact with specific molecules, known as substrates, to facilitate biochemical reactions.

Enzymes can be classified based on the types of reactions they catalyze. Common classes include:

- **Oxidoreductases**: enzymes involved in oxidation-reduction reactions (e.g., dehydrogenases, reductases).
- **Transferases**: enzymes that transfer functional groups from one molecule to another (e.g., kinases, transaminases).
- **Hydrolases**: enzymes that catalyze hydrolysis, the breaking of bonds using water (e.g., proteases, lipases).
- **Lyases**: enzymes that remove groups from molecules without hydrolysis (e.g., decarboxylases).
- **Isomerases**: enzymes that catalyze the rearrangement of molecules into isomers (e.g., epimerases).
- **Ligases**: enzymes that join two molecules together, often coupled with ATP hydrolysis (e.g., DNA ligase).

The active site of an enzyme is the region where the substrate binds. This site is highly specific to the substrate, and it facilitates the transformation of the substrate into the product(s). Enzymes may undergo conformational changes when they bind to their substrate, a process known as **induced fit**, which further enhances the catalytic process.

Enzyme-Substrate Interactions and Active Sites

The enzyme-substrate interaction follows a model called the **lock-and-key model**, where the enzyme's active site is the "lock," and the substrate is the "key" that fits perfectly. However, a more modern understanding favors the **induced fit model**, where the enzyme's active site is flexible and can mold itself to fit the substrate more effectively upon binding.

Once the substrate binds to the active site, the enzyme facilitates the conversion of the substrate to the product through various mechanisms:

1. **Proximity and Orientation**: The enzyme brings the substrate molecules closer together and aligns them in an optimal orientation for the reaction to occur.
2. **Transition State Stabilization**: The enzyme lowers the activation energy by stabilizing the transition state—the high-energy, unstable state the substrate passes through before forming the product.
3. **Microenvironment Creation**: The enzyme may create a microenvironment in its active site that is more conducive to the reaction, such as a different pH or polarity compared to the surrounding solution.
4. **Direct Participation**: In some cases, the enzyme's amino acid residues directly participate in the reaction, either donating or accepting electrons, protons, or other chemical groups.

Mechanisms of Enzyme Catalysis

Enzymes can catalyze reactions through a variety of mechanisms, including:

- **Covalent Catalysis**: In this mechanism, the enzyme forms a temporary covalent bond with the substrate during the reaction, which helps to lower the activation energy.
- **Acid-Base Catalysis**: Enzymes can donate or accept protons (H^+) during the reaction, helping to stabilize charged transition states or intermediates.
- **Metal Ion Catalysis**: Some enzymes require metal ions as cofactors, which participate directly in the catalytic process by stabilizing charged species or facilitating electron transfer.

Each of these mechanisms aids in lowering the activation energy required for the reaction to proceed, allowing biochemical reactions to occur at the rates necessary for life.

Enzyme Kinetics: Michaelis–Menten, Allosteric Regulation, and Enzyme Inhibition

Enzyme kinetics refers to the study of the rate at which enzymatic reactions occur. The most widely studied model for enzyme kinetics is the **Michaelis-Menten model**, which describes the relationship between the rate of reaction (v) and the concentration of substrate ([S]).

The Michaelis-Menten equation is expressed as:

$$v = \frac{V_{\text{max}} [S]}{K_m + [S]}$$

Where:

- **v** is the reaction velocity.
- **V_max** is the maximum velocity achieved when all enzyme active sites are saturated with substrate.
- **K_m** is the Michaelis constant, which represents the substrate concentration at which the reaction velocity is half of **V_max**. It is an indicator of the enzyme's affinity for its substrate: a low **K_m** means high affinity.

The **Michaelis-Menten constant (K_m)** and the **maximum velocity (V_max)** are important parameters in understanding how enzymes behave under different conditions. By measuring the rate of reaction at various substrate concentrations, researchers can determine these values and gain insights into enzyme function.

Allosteric Regulation

Allosteric enzymes are regulated by molecules that bind to sites other than the active site, called allosteric sites. These regulators can either activate or inhibit enzyme activity. When an allosteric effector binds to the enzyme, it induces a conformational change that either enhances or reduces the enzyme's ability to bind to the substrate.

This regulation is crucial for metabolic control, as it allows the cell to fine-tune enzyme activity in response to changing conditions. Examples of allosteric enzymes include phosphofructokinase (involved in glycolysis) and aspartate transcarbamoylase (involved in pyrimidine biosynthesis).

Enzyme Inhibition

Enzyme activity can also be controlled by inhibitors, which decrease the enzyme's catalytic efficiency. There are two main types of enzyme inhibition:

- **Competitive Inhibition**: In this type of inhibition, the inhibitor competes with the substrate for binding to the enzyme's active site. The effect of competitive inhibition can be overcome by increasing the concentration of the substrate.
- **Non-competitive Inhibition**: In non-competitive inhibition, the inhibitor binds to a site other than the active site, causing a conformational change in the enzyme that reduces its activity. This type of inhibition cannot be overcome by increasing substrate concentration.

Inhibition is a critical mechanism in regulating enzyme activity in both normal cellular processes and in therapeutic interventions, such as drug design.

Enzyme Co-factors and Co-enzymes

Many enzymes require additional non-protein molecules, called **co-factors**, to function. These can be inorganic ions, such as magnesium (Mg^{2+}) or zinc (Zn^{2+}), or organic molecules called **co-enzymes**. Co-enzymes are often derived from vitamins and assist in the enzyme's catalytic function by carrying chemical groups between molecules during reactions.

For example:

- **NAD^+** (derived from vitamin B3) serves as an electron carrier in redox reactions.
- **Coenzyme A** (derived from pantothenic acid) is involved in the transfer of acyl groups during metabolic reactions.

Enzyme cofactors and coenzymes are crucial for the proper function of many biochemical pathways, and deficiencies in these molecules can lead to metabolic diseases.

Conclusion

Enzymes are the driving forces behind nearly every biochemical reaction in the body. Their ability to catalyze reactions with precision and efficiency is fundamental to life itself. Understanding enzyme structure, function, and kinetics provides essential insights into how life processes work at the molecular level. In the next chapters, we will build on this knowledge by exploring how enzymes are involved in key metabolic pathways, such as glycolysis and the citric acid cycle, and how they contribute to energy production in cells.

Chapter 4: Metabolism: Energy for Life

Metabolism is the sum of all chemical reactions that occur within living organisms to maintain life. These reactions are essential for converting the food we eat into energy, building cellular components, and carrying out the myriad processes that sustain life. The flow of energy through metabolic pathways is tightly regulated, ensuring that cells have the energy needed to perform specific tasks while avoiding energy waste. This chapter explores the fundamental concepts of metabolism, the forms of energy in biochemical processes, and the key pathways that govern the transfer and storage of energy within the cell.

Overview of Metabolism and Energy Transfer

Metabolism can be broken down into two broad categories:

- **Catabolism**: The breakdown of molecules to release energy. Catabolic pathways typically involve the degradation of nutrients like carbohydrates, lipids, and proteins into smaller units, such as glucose, fatty acids, and amino acids. The energy released is captured in the form of high-energy molecules like ATP (adenosine triphosphate), which can then be used by the cell for various processes.
- **Anabolism**: The biosynthesis of complex molecules from simpler ones, requiring energy input. Anabolic pathways build cellular structures, such as proteins, nucleic acids, and lipids, from their respective building blocks. These pathways consume energy, typically in the form of ATP or NADPH (nicotinamide adenine dinucleotide phosphate).

Metabolism is not isolated but is a highly integrated system in which anabolic and catabolic processes are coordinated. For instance, during periods of energy excess, the body stores energy in the form of glycogen or fat, while in times of energy deficit, it mobilizes these reserves to fuel cellular processes.

ATP: The Energy Currency of the Cell

At the heart of cellular metabolism is ATP, often referred to as the "energy currency" of the cell. ATP is a nucleotide composed of adenine, ribose (a sugar), and three phosphate groups. The high-energy bonds between the phosphate groups are the key to ATP's ability to store and transfer energy.

When a cell needs energy, it breaks down ATP into ADP (adenosine diphosphate) and inorganic phosphate (Pi) in a process called **hydrolysis**. This reaction releases energy, which can be used to power various cellular activities, such as muscle contraction, active transport across membranes, and biosynthesis of macromolecules.

$$\text{ATP} + \text{H}_2\text{O} \rightarrow \text{ADP} + \text{Pi} + \text{Energy}$$

Conversely, when energy is available, ATP is synthesized from ADP and Pi, mainly through catabolic processes like cellular respiration and photosynthesis in plants.

ATP plays a crucial role in maintaining cellular energy balance and is involved in nearly every biochemical reaction. Its production is tightly controlled, and the cell ensures a steady supply of ATP to meet its energy demands.

Catabolic and Anabolic Pathways

Metabolism consists of a network of interconnected pathways that provide energy to the cell and enable the synthesis of vital molecules. These pathways can be classified into catabolic and anabolic pathways, both of which are tightly regulated to maintain metabolic balance.

Catabolic Pathways

- **Glycolysis**: The breakdown of glucose into pyruvate, producing ATP and NADH in the process. This pathway occurs in the cytoplasm and is the first step in cellular respiration.
- **Beta-Oxidation**: The breakdown of fatty acids into acetyl-CoA, which can then enter the citric acid cycle (Krebs cycle) for further energy production.
- **Proteolysis**: The breakdown of proteins into amino acids, which can be used for energy or incorporated into other biosynthetic pathways.

Anabolic Pathways

- **Gluconeogenesis**: The synthesis of glucose from non-carbohydrate precursors, ensuring that the body maintains an adequate glucose supply during periods of fasting.
- **Fatty Acid Synthesis**: The creation of fatty acids from acetyl-CoA, which are then used to form triglycerides and phospholipids, essential for energy storage and cell membrane integrity.
- **Protein Synthesis**: The assembly of amino acids into proteins, which are required for the structure, function, and regulation of the body's tissues and organs.

These pathways are governed by enzymes that are regulated by various factors, including substrate availability, feedback inhibition, and hormonal control. For instance, insulin promotes anabolic pathways like glycogenesis (the synthesis of glycogen from glucose), while glucagon stimulates catabolic pathways such as glycogenolysis (the breakdown of glycogen into glucose).

Thermodynamics in Biochemistry: Enthalpy, Entropy, and Free Energy

The study of thermodynamics in biochemistry helps explain how energy is transferred and transformed in metabolic processes. Thermodynamic principles govern the direction and spontaneity of biochemical reactions.

- **Enthalpy (ΔH)**: Enthalpy refers to the heat content of a system. In biochemical reactions, reactions that release heat are typically exothermic (ΔH < 0), while those that absorb heat are endothermic (ΔH > 0).
- **Entropy (ΔS)**: Entropy measures the disorder or randomness of a system. In general, the second law of thermodynamics states that the entropy of an isolated system tends to increase over time. Biochemical reactions that lead to an increase in entropy are more likely to occur spontaneously.
- **Free Energy (ΔG)**: The Gibbs free energy (ΔG) is the energy available to do work in a system. It combines enthalpy and entropy to predict whether a reaction will proceed spontaneously:

  $$\Delta G = \Delta H - T \Delta S$$

  where T is the temperature in Kelvin.

A negative ΔG indicates a spontaneous reaction (exergonic), meaning it can proceed without an input of energy. On the other hand, a positive ΔG indicates a non-spontaneous reaction (endergonic), which requires energy input to proceed.

Metabolic reactions can be categorized as follows:

- **Exergonic Reactions**: These reactions release energy (e.g., the breakdown of glucose in glycolysis).
- **Endergonic Reactions**: These require energy input (e.g., protein synthesis, which requires ATP).

Cells manage the flow of energy in metabolism by coupling exergonic reactions (which release energy) with endergonic reactions (which consume energy). This coupling allows the cell to perform work efficiently.

The Role of Redox Reactions in Cellular Energy Transfer

Redox reactions (reduction-oxidation reactions) play a critical role in cellular metabolism by facilitating the transfer of electrons. In these reactions, one molecule donates electrons (oxidation), while another accepts them (reduction).

In metabolic pathways like cellular respiration, redox reactions are central to the production of ATP. For example, during glycolysis and the citric acid cycle, glucose and fatty acids are oxidized, and the energy released is used to reduce electron carriers such as NAD^+ to NADH and FAD to $FADH_2$. These reduced electron carriers then transport the electrons to the electron transport chain (ETC) in mitochondria, where they are ultimately used to generate ATP through oxidative phosphorylation.

The redox reactions that occur during cellular respiration are essential for energy production and are tightly regulated to prevent damage to the cell caused by unregulated electron transfer.

Conclusion

Metabolism is the cornerstone of life, converting the energy stored in food into the forms necessary for cellular processes. The efficient flow of energy through metabolic pathways, the role of ATP as the energy currency, and the thermodynamic principles governing biochemical reactions are essential concepts in understanding how cells maintain their energy balance. In subsequent chapters, we will delve deeper into specific metabolic pathways, such as glycolysis, the citric acid cycle, and oxidative phosphorylation, to explore how energy is harnessed and utilized at the molecular level to sustain life.

Chapter 5: Glycolysis: The Breakdown of Glucose

Glycolysis is one of the most fundamental biochemical pathways in cellular metabolism. It represents the first step in the breakdown of glucose to extract energy for the cell's needs. This pathway is universally present in nearly all living organisms, from single-celled bacteria to humans, underscoring its evolutionary importance. In this chapter, we will explore the details of glycolysis, from its step-by-step breakdown to its regulation, energy yield, and the clinical implications of glycolytic enzyme deficiencies.

Step-by-Step Breakdown of Glycolysis

Glycolysis is the process by which a single molecule of glucose (a six-carbon sugar) is broken down into two molecules of pyruvate (a three-carbon compound). This pathway occurs in the cytoplasm of the cell and does not require oxygen, making it an anaerobic process. Glycolysis consists of ten enzymatically-catalyzed steps, which can be divided into two phases: the **energy investment phase** and the **energy generation phase**.

Energy Investment Phase (Steps 1–5)

In the first half of glycolysis, two ATP molecules are consumed to prepare glucose for subsequent breakdown. The steps are as follows:

1. **Hexokinase/Glucokinase**: The first step involves the phosphorylation of glucose by the enzyme hexokinase (or glucokinase in the liver). This step consumes one ATP molecule, forming glucose-6-phosphate (G6P). Phosphorylation traps glucose inside the cell since phosphorylated glucose cannot pass through the cell membrane.

2. **Phosphoglucose Isomerase**: The glucose-6-phosphate isomerizes into fructose-6-phosphate (F6P), a five-membered ring structure.

3. **Phosphofructokinase-1 (PFK-1)**: In the third step, another ATP molecule is consumed as fructose-6-phosphate is converted into fructose-1,6-bisphosphate (F1,6BP). This step is catalyzed by phosphofructokinase-1, which is the major regulatory step in glycolysis.

4. **Aldolase**: The fructose-1,6-bisphosphate is split into two three-carbon molecules: dihydroxyacetone phosphate (DHAP) and glyceraldehyde-3-phosphate (G3P).

5. **Triose Phosphate Isomerase**: The DHAP is rapidly converted into another molecule of G3P by triose phosphate isomerase. Thus, two molecules of G3P are produced from one glucose molecule.

Energy Generation Phase (Steps 6–10)

The second half of glycolysis generates energy. Each of the two molecules of G3P is processed through the following steps:

6. **Glyceraldehyde-3-Phosphate Dehydrogenase (GAPDH)**: The first step in the energy generation phase is the oxidation of G3P. The enzyme GAPDH catalyzes the transfer of two electrons and a proton to NAD+, forming NADH and producing 1,3-bisphosphoglycerate (1,3BPG).

7. **Phosphoglycerate Kinase**: 1,3-bisphosphoglycerate donates one of its high-energy phosphate groups to ADP, producing ATP and 3-phosphoglycerate (3PG) in a substrate-level phosphorylation reaction.

8. **Phosphoglycerate Mutase**: The 3-phosphoglycerate is rearranged into 2-phosphoglycerate (2PG) by phosphoglycerate mutase.

9. **Enolase**: The 2-phosphoglycerate undergoes dehydration (removal of a water molecule), resulting in the formation of phosphoenolpyruvate (PEP).

10. **Pyruvate Kinase**: In the final step of glycolysis, phosphoenolpyruvate donates its high-energy phosphate group to ADP, forming ATP and pyruvate. This step is also a substrate-level phosphorylation, and it produces two ATP molecules (one per G3P molecule).

Key Enzymes and Regulation of Glycolysis

The enzymes that catalyze the reactions of glycolysis are crucial for controlling the flow of glucose through the pathway. These enzymes are highly regulated, ensuring that glycolysis proceeds efficiently when energy is needed but is inhibited when energy levels are sufficient.

- **Hexokinase/Glucokinase**: The activity of hexokinase is inhibited by its product, glucose-6-phosphate, providing feedback regulation to prevent the overconsumption of glucose. In the liver, glucokinase has a higher Km for glucose and is regulated by insulin.

- **Phosphofructokinase-1 (PFK-1)**: This is the most important regulatory enzyme in glycolysis. PFK-1 is allosterically inhibited by high levels of ATP and citrate (an intermediate of the citric acid cycle), signaling that the cell has sufficient energy. Conversely, it is activated by AMP (which indicates low energy) and fructose-2,6-bisphosphate, a potent activator produced in response to insulin.

- **Pyruvate Kinase**: This enzyme is regulated by both allosteric effectors and covalent modification. ATP and alanine inhibit pyruvate kinase, while fructose-1,6-bisphosphate activates it, providing a mechanism for feed-forward activation in response to an abundance of substrates in earlier steps of glycolysis.

Energy Yield and Importance of Glycolysis in Cellular Metabolism

From one molecule of glucose, glycolysis generates a net gain of **2 ATP** (4 ATP are produced, but 2 are consumed in the energy investment phase) and **2 NADH**. The NADH produced will be used in the electron transport chain (in aerobic conditions) to generate additional ATP. In the absence of oxygen (anaerobic conditions), glycolysis results in the formation of lactate (in animals) or ethanol and carbon dioxide (in yeast), processes that regenerate NAD+ to allow glycolysis to continue.

Glycolysis is vital because it:

- Provides a quick source of energy, especially in tissues that require immediate ATP, such as muscle during exercise.
- Forms the starting point for aerobic respiration (via pyruvate entry into the citric acid cycle) or anaerobic fermentation.
- Is the primary metabolic pathway in tissues without mitochondria (e.g., red blood cells).
- Plays a role in the biosynthesis of intermediates required for other metabolic pathways (e.g., amino acids, nucleotides).

Glycolysis in Different Tissues and Organisms

Glycolysis occurs in nearly all cells, but its regulation can vary between tissues depending on their specific metabolic demands.

- **Muscle cells**: During intense exercise, muscle cells rely heavily on glycolysis for rapid ATP production in the absence of sufficient oxygen, leading to the formation of lactate (lactic acid fermentation). This process is reversible, as lactate can later be converted back to glucose in the liver via the Cori cycle.
- **Red blood cells**: Since red blood cells lack mitochondria, they depend entirely on glycolysis for ATP production. The absence of mitochondria also means that they cannot utilize oxidative phosphorylation for energy, making glycolysis their sole energy pathway.
- **Liver cells**: In the liver, glycolysis is tightly regulated to maintain blood glucose levels. It is activated by insulin following meals (when glucose levels are high) and inhibited by glucagon and epinephrine during fasting (when glucose levels are low).
- **Yeast and bacteria**: These organisms use glycolysis as part of fermentation processes to generate energy in the absence of oxygen. Yeast produces ethanol and CO_2 as byproducts of fermentation, which is exploited in the production of alcoholic beverages and bread.

Disorders Related to Glycolytic Enzyme Deficiencies

Several metabolic disorders are associated with deficiencies in glycolytic enzymes. These deficiencies often result in impaired energy production, leading to a variety of clinical symptoms.

- **Pyruvate Kinase Deficiency**: This is one of the most common glycolytic enzyme deficiencies, particularly in red blood cells. It leads to hemolytic anemia because the lack of ATP in red blood cells impairs their function and leads to premature cell death.
- **Phosphofructokinase Deficiency**: This rare disorder leads to muscle cramping, weakness, and myopathy, especially after exercise. The inability to properly regulate glycolysis during muscle activity impairs energy production.
- **Hexokinase Deficiency**: This disorder is extremely rare and typically results in early-onset hemolytic anemia, as the inability to phosphorylate glucose prevents its use in glycolysis.

These deficiencies highlight the critical role glycolysis plays in energy production, especially in energy-demanding tissues like red blood cells and muscle.

Conclusion

Glycolysis is an essential metabolic pathway that plays a central role in cellular energy production. Understanding its mechanics, regulation, and importance in various tissues helps us appreciate its role in maintaining cellular function and homeostasis. The next chapter will delve into the citric acid cycle (Krebs cycle), where the products of glycolysis are further metabolized to produce a significant amount of ATP.

Chapter 6: The Citric Acid Cycle (Krebs Cycle)

The Citric Acid Cycle, also known as the Krebs cycle or the TCA cycle (tricarboxylic acid cycle), is one of the central pathways in cellular metabolism. It plays a critical role in the production of energy by oxidizing acetyl-CoA, derived from carbohydrates, fats, and proteins, into carbon dioxide and high-energy electrons. These electrons are later used to produce ATP, the cell's primary energy currency. This chapter will break down the reactions, regulation, and significance of the citric acid cycle in metabolism.

Overview and Significance of the Citric Acid Cycle

The citric acid cycle takes place in the mitochondria of eukaryotic cells and is a key component of both aerobic respiration and metabolism. It serves as a hub, linking carbohydrate, lipid, and protein metabolism, and provides the cell with essential intermediates for the synthesis of macromolecules and other metabolic processes. The main goal of the cycle is to generate high-energy electron carriers (NADH and $FADH_2$), which will be used in oxidative phosphorylation to produce ATP.

In addition to energy production, the citric acid cycle provides intermediates for biosynthetic pathways. These intermediates are used for the synthesis of amino acids, nucleotides, and other compounds essential for cell function and growth.

Reactions of the Krebs Cycle and Key Intermediates

The citric acid cycle consists of a series of eight enzyme-catalyzed reactions. Here's a step-by-step breakdown:

Acetyl-CoA Formation and Entry into the Cycle

The cycle begins with the entry of acetyl-CoA into the mitochondria. Acetyl-CoA is typically derived from pyruvate (the product of glycolysis), fatty acids, or certain amino acids.

Citrate Formation

Acetyl-CoA (a two-carbon molecule) reacts with oxaloacetate (a four-carbon molecule) to form citrate (a six-carbon molecule). This reaction is catalyzed by citrate synthase.

Isomerization of Citrate

Citrate undergoes isomerization to form isocitrate, catalyzed by the enzyme aconitase. This step involves the dehydration and rehydration of citrate to produce isocitrate.

Oxidative Decarboxylation to Alpha-Ketoglutarate

Isocitrate is oxidized by NAD+ to form oxalosuccinate, a short-lived intermediate. Oxalosuccinate is then decarboxylated to form alpha-ketoglutarate, a five-carbon compound. This step produces NADH and releases CO_2.

Formation of Succinyl-CoA

Alpha-ketoglutarate undergoes a second oxidative decarboxylation, catalyzed by the enzyme alpha-ketoglutarate dehydrogenase, resulting in the formation of succinyl-CoA, a four-carbon compound. This reaction also produces NADH and releases another molecule of CO_2.

Conversion of Succinyl-CoA to Succinate

Succinyl-CoA is converted into succinate by the enzyme succinyl-CoA synthetase. This reaction generates GTP (or ATP, depending on the tissue), which is an energy carrier, and also releases CoA.

Oxidation of Succinate to Fumarate

Succinate is oxidized to fumarate by succinate dehydrogenase, producing $FADH_2$, another high-energy electron carrier. This step occurs on the inner mitochondrial membrane, as succinate dehydrogenase is part of the electron transport chain.

Hydration of Fumarate to Malate

Fumarate is hydrated to form malate, catalyzed by fumarase. This reaction involves the addition of a water molecule to fumarate.

Oxidation of Malate to Oxaloacetate

Finally, malate is oxidized to regenerate oxaloacetate, the starting molecule of the cycle, by malate dehydrogenase. This step generates NADH.

At this point, oxaloacetate is ready to combine with a new molecule of acetyl-CoA, starting the cycle again. For every turn of the citric acid cycle, two molecules of CO2 are released, along with three molecules of NADH, one molecule of FADH2, and one molecule of GTP (or ATP).

Regulation and Control of the Citric Acid Cycle

The citric acid cycle is tightly regulated at key steps to ensure that energy production is matched to the cell's needs. The primary regulatory points are:

- **Citrate Synthase**: The enzyme that catalyzes the formation of citrate from acetyl-CoA and oxaloacetate is inhibited by high concentrations of ATP, NADH, and succinyl-CoA, signaling that the cell's energy needs are met.
- **Isocitrate Dehydrogenase**: This enzyme is regulated by ATP and NADH, which inhibit it when energy levels are high. Conversely, ADP and NAD+ activate it, signaling a need for energy production.
- **Alpha-Ketoglutarate Dehydrogenase**: This enzyme is also regulated by feedback inhibition from products like NADH and succinyl-CoA. When energy is abundant, the cycle slows down to conserve resources.

In addition to these points of feedback inhibition, the cycle can be influenced by the availability of acetyl-CoA, NAD+, and oxygen, as the cycle is part of aerobic respiration and requires oxygen to function optimally.

Connections to Other Metabolic Pathways

The citric acid cycle is not an isolated pathway; it is interconnected with several other metabolic pathways:

- **Glycolysis**: Pyruvate, the end product of glycolysis, is converted to acetyl-CoA before entering the citric acid cycle.
- **Fatty Acid Metabolism**: Fatty acids are broken down into two-carbon units as acetyl-CoA, which enters the citric acid cycle.
- **Amino Acid Catabolism**: Certain amino acids can be converted into intermediates of the citric acid cycle, allowing for integration between protein and carbohydrate/fat metabolism.

Additionally, intermediates of the cycle are used in the biosynthesis of amino acids, glucose (via gluconeogenesis), nucleotides, and heme, demonstrating the cycle's central role in metabolism.

Diseases Associated with Krebs Cycle Enzyme Deficiencies

Deficiencies or dysfunctions in enzymes involved in the citric acid cycle can lead to serious metabolic disorders. Some notable conditions include:

- **Alpha-Ketoglutarate Dehydrogenase Deficiency**: This condition results in neurological impairment and developmental delay, as alpha-ketoglutarate is crucial for neurotransmitter synthesis.
- **Fumarase Deficiency**: A rare genetic disorder characterized by developmental delay, seizures, and other neurological impairments.
- **Succinic Semialdehyde Dehydrogenase Deficiency**: This metabolic disorder leads to the accumulation of succinic semialdehyde, causing neurodevelopmental and psychiatric symptoms.

These conditions highlight the importance of the citric acid cycle in maintaining normal metabolic function and the consequences of its disruption.

Conclusion

The citric acid cycle is a cornerstone of cellular metabolism, enabling cells to extract energy from nutrients and provide intermediates for essential biosynthetic processes. Through its regulation and integration with other metabolic pathways, it ensures that the cell can meet its energy demands while maintaining metabolic balance. Understanding the intricacies of the citric acid cycle not only sheds light on fundamental cellular processes but also provides insights into various metabolic diseases, highlighting the critical need for maintaining healthy mitochondrial function in cellular health.

Chapter 7: Oxidative Phosphorylation and the Electron Transport Chain

Oxidative phosphorylation, the final stage of cellular respiration, is the process by which cells generate the majority of their ATP. This chapter will explore the critical role of the electron transport chain (ETC) and how it drives ATP synthesis in the mitochondria. By examining the components, mechanisms, and regulation of this process, we will uncover how cells harness energy from the oxidation of nutrients and convert it into usable cellular energy.

Structure and Function of Mitochondria in Energy Production

Mitochondria are often referred to as the "powerhouses" of the cell because they are the sites where oxidative phosphorylation occurs. These organelles are double-membraned, with an outer membrane and a highly invaginated inner membrane. The inner membrane contains the components of the electron transport chain and the ATP synthase machinery, while the matrix (the space inside the inner membrane) houses the enzymes for the citric acid cycle and fatty acid oxidation.

The structure of mitochondria is perfectly designed for energy production:

- The **outer membrane** is relatively permeable to small molecules, allowing the passage of ions and metabolites.
- The **inner membrane** is highly selective and contains the proteins necessary for the electron transport chain and ATP synthesis. Its extensive surface area, due to folds called **cristae**, maximizes the capacity for ATP production.

Mitochondria also have their own DNA, enabling them to produce some of their own proteins involved in energy metabolism. However, the majority of mitochondrial proteins are encoded by nuclear DNA and imported into the mitochondria.

The Electron Transport Chain (ETC): Components and Function

The electron transport chain is a series of protein complexes located in the inner mitochondrial membrane. These complexes transfer electrons derived from nutrients (primarily NADH and FADH2) to molecular oxygen (O_2), the final electron acceptor. The energy released during these electron transfers is used to pump protons (H^+) across the inner mitochondrial membrane, creating a proton gradient.

The main components of the electron transport chain include:

1. **Complex I (NADH Dehydrogenase)**: This complex accepts electrons from NADH and passes them to ubiquinone (CoQ). During this process, protons are pumped from the matrix into the intermembrane space.

2. **Complex II (Succinate Dehydrogenase)**: Succinate, produced in the citric acid cycle, donates electrons to ubiquinone through complex II, generating FADH2 in the process. Unlike complex I, complex II does not pump protons.

3. **Ubiquinone (Coenzyme Q)**: Ubiquinone is a mobile electron carrier that transfers electrons from complexes I and II to complex III. It also helps shuttle protons across the membrane.

4. **Complex III (Cytochrome bc1 Complex)**: Complex III receives electrons from ubiquinone and transfers them to cytochrome c. As electrons move through complex III, more protons are pumped into the intermembrane space.

5. **Cytochrome c**: Cytochrome c is a small heme-containing protein that acts as an electron carrier between complex III and complex IV. It is loosely attached to the inner mitochondrial membrane.

6. **Complex IV (Cytochrome c Oxidase)**: This complex receives electrons from cytochrome c and passes them to molecular oxygen (O_2), reducing it to water (H_2O). This step is crucial as it consumes oxygen, completing the electron transport chain.

7. **ATP Synthase**: ATP synthase is a protein complex located in the inner mitochondrial membrane. It is not part of the electron transport chain itself, but it utilizes the proton gradient created by the ETC to synthesize ATP from ADP and inorganic phosphate (Pi). As protons flow back into the mitochondrial matrix through ATP synthase, the energy released is used to phosphorylate ADP, forming ATP.

Proton Gradient and ATP Synthesis in Oxidative Phosphorylation

The primary goal of the electron transport chain is to establish a **proton gradient** across the inner mitochondrial membrane. As electrons are transferred through the protein complexes, protons are actively pumped from the mitochondrial matrix into the intermembrane space. This creates an electrochemical gradient, often referred to as the **proton motive force (PMF)**.

The proton motive force consists of two components:

1. **The concentration gradient of protons** (higher proton concentration in the intermembrane space compared to the matrix).
2. **The electrical gradient** (a difference in charge between the intermembrane space and the matrix).

The proton gradient is essential for ATP production. Protons flow back into the matrix through ATP synthase, which harnesses this flow to drive the phosphorylation of ADP into ATP. This process is known as **chemiosmosis**. The energy generated from the proton flow is what powers the ATP synthesis.

Role of Oxygen in Cellular Respiration

Oxygen plays a critical role in the electron transport chain as the final electron acceptor. Without oxygen, electrons would accumulate in the protein complexes of the electron transport chain, halting the entire process. Oxygen binds to the electrons at complex IV and combines with protons to form water. This step is essential for maintaining the flow of electrons through the chain and sustaining ATP production.

Because of its role as the terminal electron acceptor, oxygen is crucial to aerobic respiration. In the absence of oxygen, cells must rely on anaerobic pathways, such as fermentation, to generate ATP, but these processes are far less efficient in terms of energy yield.

Pathologies Related to Oxidative Phosphorylation

Dysfunction in the electron transport chain can lead to a variety of serious metabolic and genetic disorders. Some notable conditions include:

1. **Mitochondrial Diseases**: These diseases are caused by mutations in mitochondrial DNA or nuclear genes encoding mitochondrial proteins. Such mutations can impair electron transport chain function, leading to insufficient ATP production. Mitochondrial diseases often affect tissues with high energy demands, such as the brain, heart, and muscles.
2. **Leber's Hereditary Optic Neuropathy (LHON)**: A genetic disorder that leads to optic nerve degeneration and blindness, often caused by mutations in genes encoding subunits of complex I.
3. **Parkinson's Disease**: Some forms of Parkinson's disease are associated with dysfunction in complex I, leading to a reduction in ATP production and increased oxidative stress in neurons.
4. **Alzheimer's Disease**: Reduced mitochondrial function and energy production in the brain are implicated in the pathogenesis of Alzheimer's disease. Mitochondrial dysfunction can lead to neuronal death and contribute to cognitive decline.
5. **Ischemia and Hypoxia**: In conditions where oxygen supply is limited, such as during a heart attack or stroke, the electron transport chain becomes impaired, leading to a lack of ATP production and contributing to tissue damage.

Conclusion

Oxidative phosphorylation and the electron transport chain are central to cellular energy production. By transferring electrons through a series of protein complexes and using the resulting proton gradient to generate ATP, these processes provide the cell with the energy required for essential functions. The regulation of oxidative phosphorylation is tightly coupled with cellular metabolism, ensuring energy production is aligned with the cell's needs. Dysfunction in any part of this system can have severe consequences, highlighting the critical role mitochondria play in maintaining cellular health. Understanding oxidative phosphorylation opens up new possibilities in treating diseases related to mitochondrial dysfunction and aging, making it a vital area of biochemistry and medical research.

Chapter 8: Gluconeogenesis and Glycogen Metabolism

In this chapter, we will explore the biochemical processes responsible for the synthesis and storage of glucose, essential for maintaining energy homeostasis in the body. Gluconeogenesis and glycogen metabolism are key components of metabolic regulation, especially during periods of fasting, exercise, or stress. These processes ensure a steady supply of glucose, the primary energy source for the brain and other tissues, even when dietary glucose is not available.

Gluconeogenesis: The Synthesis of Glucose from Non-Carbohydrate Sources

Gluconeogenesis is the process by which the body synthesizes glucose from non-carbohydrate precursors, such as lactate, glycerol, and certain amino acids. This pathway primarily occurs in the liver and, to a lesser extent, in the kidneys. It is essentially the reverse of glycolysis, though it bypasses some of the irreversible steps in glycolysis.

Key steps and enzymes in gluconeogenesis:

1. **Pyruvate Carboxylase**: This enzyme catalyzes the conversion of pyruvate to oxaloacetate in the mitochondria. This is the first committed step in gluconeogenesis.
2. **Phosphoenolpyruvate Carboxykinase (PEPCK)**: Oxaloacetate is decarboxylated and phosphorylated to form phosphoenolpyruvate (PEP), which is the key intermediate in gluconeogenesis.
3. **Fructose-1,6-Bisphosphatase**: This enzyme bypasses the irreversible step of phosphofructokinase in glycolysis by converting fructose-1,6-bisphosphate to fructose-6-phosphate.
4. **Glucose-6-Phosphatase**: This enzyme removes the phosphate group from glucose-6-phosphate to produce free glucose, which is then released into the bloodstream.

These enzymatic steps are regulated by hormonal signals such as insulin, glucagon, and cortisol, which ensure that glucose is synthesized when the body needs it, and suppressed when glucose is abundant.

Key Enzymes and Regulatory Mechanisms in Gluconeogenesis

Gluconeogenesis is tightly regulated to balance glucose production with the body's energy needs. The regulation occurs through both hormonal control and allosteric modulation of enzymes:

Hormonal Regulation

- **Insulin**: Lowers blood glucose levels and inhibits gluconeogenesis by deactivating key enzymes like fructose-1,6-bisphosphatase and PEPCK.
- **Glucagon**: Stimulates gluconeogenesis by activating key enzymes. In times of fasting or low blood glucose, glucagon increases the activity of pyruvate carboxylase and PEPCK, promoting glucose synthesis.
- **Cortisol**: This stress hormone stimulates gluconeogenesis by increasing enzyme expression, helping to maintain glucose levels during periods of stress or prolonged fasting.
- **Epinephrine**: Stimulates gluconeogenesis to provide glucose for quick energy in response to stress or exercise.

Allosteric Regulation

- **Acetyl-CoA**: Acts as an allosteric activator of pyruvate carboxylase, ensuring that the pathway can proceed when there is a sufficient supply of energy.
- **AMP**: AMP is a signal of low energy, and its presence inhibits key enzymes like fructose-1,6-bisphosphatase, reducing gluconeogenesis when energy is scarce.

Gluconeogenesis serves as an essential adaptive response, providing glucose to tissues like the brain that rely heavily on it, even during fasting or when dietary intake is insufficient.

Glycogen Storage and Mobilization

Glycogen is the storage form of glucose, predominantly found in the liver and muscles. It acts as a readily available source of glucose when the body requires a quick energy boost. Glycogen metabolism involves both the **synthesis** and **breakdown** of glycogen, processes regulated by hormones and allosteric factors.

Glycogen Synthesis (Glycogenesis)

- **Glycogen Synthase**: The key enzyme in glycogenesis catalyzes the addition of glucose units to a growing glycogen chain. It uses **UDP-glucose** as the donor of glucose residues.
- **Branching Enzyme**: This enzyme creates branches in the glycogen molecule, allowing for a more compact structure that can be rapidly mobilized.
- **Regulation**: Insulin promotes glycogenesis by activating glycogen synthase and inhibiting glycogen phosphorylase (the enzyme involved in glycogen breakdown).

Glycogen Breakdown (Glycogenolysis)

- **Glycogen Phosphorylase**: The enzyme responsible for breaking down glycogen into glucose-1-phosphate. It is activated by glucagon and epinephrine during periods of low blood glucose or during physical exertion.
- **Debranching Enzyme**: This enzyme removes branches from the glycogen molecule to allow for continued breakdown.
- **Regulation**: Glucagon and epinephrine trigger glycogenolysis through the activation of glycogen phosphorylase. Conversely, insulin inhibits glycogen breakdown by deactivating glycogen phosphorylase.

Glycogen is an essential energy reserve that helps maintain blood glucose levels between meals and provides rapid glucose availability during exercise.

Diseases of Glycogen Metabolism: Glycogen Storage Diseases

Disruptions in glycogen metabolism can lead to **glycogen storage diseases (GSDs)**, a group of inherited disorders that affect the enzymes involved in glycogen synthesis or breakdown. These diseases can result in abnormal glycogen accumulation in tissues like the liver and muscles, causing a range of symptoms depending on the specific enzyme defect.

- **GSD Type I (von Gierke Disease)**: Caused by a deficiency in glucose-6-phosphatase, this disorder impairs the ability of the liver to release glucose into the bloodstream, leading to hypoglycemia (low blood sugar). Affected individuals may experience growth failure, enlarged liver (hepatomegaly), and lactic acidosis.
- **GSD Type II (Pompe Disease)**: Caused by a deficiency in **acid alpha-glucosidase**, an enzyme involved in glycogen breakdown within lysosomes. This leads to glycogen accumulation in the heart and muscles, causing muscle weakness and respiratory issues.
- **GSD Type V (McArdle Disease)**: Caused by a deficiency in **muscle glycogen phosphorylase**, this disease affects the breakdown of glycogen in muscles, leading to exercise intolerance, muscle cramps, and myoglobinuria (the presence of muscle proteins in the urine after exertion).

These diseases underscore the importance of properly regulating glycogen synthesis and breakdown, as disturbances can lead to significant metabolic dysfunction.

Hormonal Regulation of Glucose and Glycogen Metabolism

The balance between glucose production, storage, and mobilization is tightly controlled by hormonal signals, primarily **insulin** and **glucagon**:

- **Insulin**: The primary hormone responsible for lowering blood glucose levels. It promotes glycogenesis, inhibits gluconeogenesis, and suppresses glycogenolysis.
- **Glucagon**: Released when blood glucose levels are low, glucagon stimulates gluconeogenesis and glycogenolysis to increase glucose availability.
- **Epinephrine**: During periods of stress or exercise, epinephrine stimulates glycogen breakdown to provide immediate energy.
- **Cortisol**: This hormone aids in maintaining blood glucose levels by promoting gluconeogenesis, particularly during prolonged fasting or stress.

This hormonal interplay ensures that glucose is available when needed, whether for immediate energy or for long-term storage.

Conclusion

Gluconeogenesis and glycogen metabolism are critical processes that help maintain glucose homeostasis in the body. Gluconeogenesis allows for the synthesis of glucose from non-carbohydrate precursors, while glycogen metabolism provides a rapid source of glucose when needed. Both processes are tightly regulated by enzymes and hormones to ensure that the body has a constant and adequate supply of glucose, supporting energy demands and overall metabolic balance. Understanding these pathways is fundamental not only for biochemistry but also for understanding diseases like diabetes and glycogen storage disorders.

Chapter 9: Lipid Metabolism: Fatty Acids and Beyond

Lipid metabolism plays a central role in energy production, cellular structure, and signaling within the body. Unlike carbohydrates, which are the body's primary energy source, lipids provide a dense form of energy storage. Fatty acids, triglycerides, phospholipids, and cholesterol are some of the key components involved in lipid metabolism. This chapter will explore the pathways of lipid breakdown and synthesis, their regulatory mechanisms, and the disorders that can arise from disturbances in lipid metabolism.

9.1 Lipid Classification and Functions

Lipids are a diverse group of molecules that are hydrophobic and play a variety of essential roles in biological systems. They can be classified into several categories:

- **Fatty Acids**: Long chains of hydrocarbons that serve as building blocks for many other lipids. They can be saturated (no double bonds) or unsaturated (one or more double bonds).
- **Triglycerides**: The primary form of stored energy in the body, consisting of three fatty acids bound to a glycerol molecule.
- **Phospholipids**: Key components of cell membranes, consisting of a glycerol backbone, two fatty acid chains, and a phosphate group.
- **Cholesterol**: A type of sterol, important for membrane structure and as a precursor for the synthesis of steroid hormones and bile acids.

Lipids are crucial not only for energy storage but also for maintaining cellular integrity and communication. They provide structure to cell membranes, enable the storage of energy in adipose tissue, and play vital roles in signaling pathways.

9.2 Breakdown of Fatty Acids: Beta-Oxidation

The breakdown of fatty acids begins with **beta-oxidation**, a process that occurs in the mitochondria of cells. Here, fatty acids are converted into acetyl-CoA, which enters the citric acid cycle (Krebs cycle) to produce ATP. This process occurs in several steps:

1. **Activation**: Fatty acids are first activated in the cytoplasm by binding to coenzyme A (CoA), forming fatty acyl-CoA.
2. **Transport**: The fatty acyl-CoA is then transported into the mitochondria via the carnitine shuttle.
3. **Beta-Oxidation**: Once inside the mitochondria, the fatty acid undergoes repeated cycles of oxidation, producing acetyl-CoA, NADH, and FADH2.

Each cycle shortens the fatty acid chain by two carbon atoms, and the acetyl-CoA produced enters the citric acid cycle to be further oxidized for energy. The NADH and FADH2 generated are used in the electron transport chain for ATP production.

9.3 Synthesis of Fatty Acids: Lipogenesis

Fatty acid synthesis, or **lipogenesis**, is the process by which the body converts excess carbohydrates and proteins into fatty acids. This occurs primarily in the liver and adipose tissue and is regulated by enzymes such as **acetyl-CoA carboxylase** and **fatty acid synthase**. The key steps include:

1. **Acetyl-CoA Formation**: Acetyl-CoA is derived from glucose metabolism via glycolysis.
2. **Malonyl-CoA Formation**: Acetyl-CoA is converted to malonyl-CoA, which is an essential intermediate in fatty acid synthesis.
3. **Fatty Acid Elongation**: Malonyl-CoA is used by fatty acid synthase to elongate the carbon chain, adding two carbon atoms at a time.

The synthesized fatty acids can be converted into triglycerides for storage or incorporated into phospholipids for membrane construction.

9.4 Lipoprotein Metabolism

Lipoproteins are complexes of lipids and proteins that transport lipids through the bloodstream. The main types of lipoproteins include:

- **Chylomicrons**: Formed in the intestine, they transport dietary triglycerides and cholesterol to peripheral tissues.
- **Very Low-Density Lipoproteins (VLDL)**: Produced by the liver, they carry triglycerides and cholesterol to tissues.
- **Low-Density Lipoproteins (LDL)**: Known as "bad cholesterol," LDL transports cholesterol to peripheral tissues. Elevated levels of LDL are associated with an increased risk of cardiovascular diseases.
- **High-Density Lipoproteins (HDL)**: Known as "good cholesterol," HDL removes excess cholesterol from tissues and transports it to the liver for excretion.

Lipoproteins play an essential role in lipid metabolism by facilitating the transport of fats in a water-soluble form, ensuring their proper distribution and utilization in the body.

9.5 Ketogenesis and Ketone Bodies

In conditions where glucose availability is low, such as fasting or prolonged exercise, the liver produces **ketone bodies** through a process called **ketogenesis**. Ketone bodies (acetoacetate, beta-hydroxybutyrate, and acetone) are produced from acetyl-CoA in the liver and serve as an alternative fuel source for tissues, particularly the brain and muscles.

The production of ketone bodies occurs in the mitochondria of liver cells, where excess acetyl-CoA from fatty acid oxidation is converted into ketone bodies. These ketones are then released into the bloodstream and transported to other tissues, where they are converted back into acetyl-CoA for energy production.

Ketogenesis is a vital adaptation to prolonged periods of fasting, carbohydrate restriction, or intense physical activity, allowing the body to maintain energy production when glucose is scarce.

9.6 Disorders of Lipid Metabolism

Several inherited and acquired disorders can disrupt lipid metabolism, leading to a range of health problems. Some of the most common disorders include:

- **Familial Hypercholesterolemia**: A genetic disorder characterized by high levels of LDL cholesterol, leading to an increased risk of cardiovascular disease.
- **Gaucher's Disease**: A rare genetic disorder caused by a deficiency in the enzyme glucocerebrosidase, leading to the accumulation of lipids in various tissues, particularly the spleen and liver.
- **Lipodystrophy**: A group of disorders where there is either a loss or abnormal distribution of body fat, often associated with metabolic abnormalities such as insulin resistance and hypertriglyceridemia.
- **Sphingolipidoses**: A group of disorders, such as Tay-Sachs disease, in which the accumulation of sphingolipids disrupts normal cell function, particularly in the nervous system.

Disruptions in lipid metabolism can lead to serious health complications, including cardiovascular disease, liver dysfunction, and neurological disorders.

9.7 Hormonal Regulation of Lipid Metabolism

Lipid metabolism is tightly regulated by hormones that respond to changes in nutrient availability, energy demands, and metabolic state. Key hormones involved in lipid metabolism include:

- **Insulin**: Stimulates the storage of fatty acids in adipose tissue and promotes lipogenesis while inhibiting lipolysis.
- **Glucagon**: Stimulates the breakdown of stored triglycerides in adipose tissue (lipolysis) and promotes fatty acid release into the bloodstream for energy production.
- **Epinephrine**: Also stimulates lipolysis, particularly during stress or physical activity.
- **Cortisol**: A glucocorticoid hormone that promotes the mobilization of fatty acids from adipose tissue during periods of fasting or stress.

The balance between these hormones ensures that lipid metabolism adapts to the body's energy needs, facilitating the efficient storage and utilization of fats.

9.8 Conclusion

Lipid metabolism is a complex and vital process that supports energy production, cellular structure, and signaling in the body. From the breakdown of fatty acids for energy to the synthesis of lipids for cellular membranes, the body relies on intricate biochemical pathways to maintain balance. Understanding lipid metabolism provides insights into the mechanisms of energy regulation, the pathophysiology of metabolic disorders, and the potential for therapeutic interventions targeting lipid-related diseases. As we continue to uncover the molecular details of lipid metabolism, new strategies for treating disorders such as cardiovascular disease, obesity, and diabetes may emerge, offering hope for improved health outcomes.

Chapter 10: The Urea Cycle and Nitrogen Metabolism

The urea cycle (also known as the ornithine cycle) plays a critical role in maintaining nitrogen balance and preventing the toxic accumulation of ammonia in the body. Nitrogen, primarily obtained from the metabolism of amino acids, is a vital element for building proteins and other cellular components. However, excess nitrogen must be efficiently processed and excreted to maintain cellular homeostasis. This chapter will explore the biochemical processes involved in nitrogen metabolism, focusing on the urea cycle, amino acid catabolism, and disorders related to nitrogen metabolism.

10.1 The Urea Cycle: Ammonia Detoxification and Nitrogen Excretion

The urea cycle is the primary mechanism for detoxifying ammonia, a byproduct of amino acid catabolism. Ammonia is highly toxic, especially to the brain, and its accumulation in the bloodstream can lead to a condition known as **hyperammonemia**, which can cause severe neurological symptoms. The urea cycle, which occurs in the liver, converts ammonia into urea, a less toxic compound that can be safely excreted by the kidneys.

The cycle consists of a series of enzymatic reactions that convert two molecules of ammonia into urea, which is then transported to the kidneys for excretion in urine. The key steps of the urea cycle are as follows:

1. **Ammonia Fixation**: Ammonia is first combined with carbon dioxide to form carbamoyl phosphate in a reaction catalyzed by the enzyme **carbamoyl phosphate synthetase I**.
2. **Citrulline Formation**: Carbamoyl phosphate reacts with ornithine to form **citrulline**, which is then transported from the mitochondria to the cytoplasm.
3. **Argininosuccinate Synthesis**: Citrulline reacts with aspartate to form **argininosuccinate**, a key intermediate in the cycle, through the enzyme **argininosuccinate synthetase**.
4. **Fumarate Production**: Argininosuccinate is cleaved by **argininosuccinate lyase** to produce **fumarate** and **arginine**.
5. **Urea Formation**: Arginine is hydrolyzed by **arginase** to form **urea** and regenerate ornithine, which can re-enter the cycle.

The urea cycle thus efficiently converts toxic ammonia into urea, which is then safely excreted by the kidneys. This cycle is essential for maintaining nitrogen balance in the body, especially during periods of protein catabolism.

10.2 Amino Acid Catabolism and Nitrogen Balance

Amino acids are the building blocks of proteins, and their metabolism is central to maintaining nitrogen balance in the body. When amino acids are broken down, they release nitrogen in the form of ammonia. This ammonia must be processed to prevent toxicity. The nitrogen removed from amino acids during their breakdown is either converted to urea via the urea cycle or excreted in other ways.

The main steps in amino acid catabolism include:

1. **Transamination**: The first step in amino acid degradation is the transfer of the amino group (−NH2) to a keto acid, typically **alpha-ketoglutarate**, which forms **glutamate**. This process is catalyzed by **aminotransferases**.
2. **Deamination**: In the next step, glutamate undergoes oxidative deamination catalyzed by **glutamate dehydrogenase**, releasing ammonia and converting glutamate into alpha-ketoglutarate. The ammonia produced is then processed by the urea cycle.
3. **Amino Acid-Specific Pathways**: Each amino acid follows specific pathways for its catabolism. For example, **branched-chain amino acids** (leucine, isoleucine, and valine) are primarily catabolized in muscle tissue, while other amino acids like **glutamine** are important for transporting ammonia between tissues.

Nitrogen balance refers to the balance between nitrogen intake (mainly through dietary protein) and nitrogen excretion. In a healthy adult, nitrogen intake and excretion are balanced, meaning the body neither accumulates nor loses nitrogen. Disruptions in nitrogen metabolism, such as in liver dysfunction, can lead to imbalances and result in excess ammonia in the bloodstream.

10.3 Disorders of Nitrogen Metabolism

Disorders of nitrogen metabolism can lead to severe health problems, primarily due to the accumulation of toxic ammonia. Some of the most notable disorders related to nitrogen metabolism include:

- **Hyperammonemia**: Hyperammonemia is a condition characterized by elevated levels of ammonia in the bloodstream. It can result from defects in the enzymes of the urea cycle, such as **carbamoyl phosphate synthetase I**, **ornithine transcarbamylase**, or **arginase**. Symptoms of hyperammonemia include vomiting, lethargy, and in severe cases, encephalopathy or coma.
- **Urea Cycle Defects**: Inherited defects in the enzymes involved in the urea cycle can lead to a buildup of ammonia. For example, **ornithine transcarbamylase deficiency** is the most common urea cycle disorder, and it can cause severe hyperammonemia. Early diagnosis and treatment are crucial for preventing irreversible brain damage in affected infants.
- **Citrullinemia**: This rare genetic disorder is caused by a deficiency of the enzyme **argininosuccinate synthetase**, leading to an accumulation of citrulline and ammonia. Symptoms include vomiting, lethargy, and neurological impairment.
- **Phenylketonuria (PKU)**: Although not directly related to the urea cycle, PKU is a metabolic disorder that impairs the catabolism of the amino acid phenylalanine. The accumulation of phenylalanine can lead to intellectual disability and other neurological issues. PKU is typically managed by dietary restriction of phenylalanine.

- **Liver Disease and Ammonia Toxicity**: Chronic liver diseases, such as cirrhosis, can impair the urea cycle, leading to impaired detoxification of ammonia. In such cases, ammonia builds up in the bloodstream, causing **hepatic encephalopathy**, a potentially fatal condition characterized by confusion, altered consciousness, and coma.

10.4 The Role of the Liver and Kidneys in Nitrogen Waste Elimination

The **liver** and **kidneys** are the primary organs involved in nitrogen metabolism and waste elimination:

- **Liver**: The liver is the central site for the urea cycle, where ammonia is converted into urea. The liver also synthesizes glutamine, an important molecule for ammonia detoxification, and regulates amino acid catabolism.
- **Kidneys**: The kidneys play a crucial role in eliminating urea from the bloodstream. Urea is filtered by the kidneys and excreted in urine. In cases of kidney dysfunction, the ability to excrete nitrogen waste is compromised, leading to uremia (elevated urea levels in the blood).

Both organs are integral to maintaining nitrogen homeostasis, and dysfunction in either can have profound effects on the body's ability to detoxify ammonia and maintain proper nitrogen balance.

10.5 Conclusion

The urea cycle and nitrogen metabolism are fundamental to the body's ability to process and excrete excess nitrogen derived from amino acid catabolism. Proper functioning of these processes is essential for maintaining metabolic homeostasis and preventing the accumulation of toxic waste products like ammonia. Disruptions in nitrogen metabolism, whether through genetic defects, liver disease, or kidney dysfunction, can lead to serious clinical conditions, highlighting the importance of these metabolic pathways in health. Understanding these biochemical processes provides critical insights into the treatment of metabolic disorders and underscores the importance of liver and kidney health in maintaining overall well-being.

Chapter 11: Protein Metabolism and Amino Acid Synthesis

Proteins are essential macromolecules in every cell, and their metabolism is crucial for maintaining cellular function, structure, and overall health. Protein metabolism involves the synthesis, turnover, and degradation of proteins, as well as the synthesis of amino acids, which are the building blocks for protein production. This chapter explores the biochemical pathways involved in protein metabolism and amino acid synthesis, their regulation, and the disorders associated with these processes.

11.1 Overview of Protein Turnover and Degradation

Protein turnover is the continuous process by which proteins are synthesized and degraded within cells. This process ensures that damaged or misfolded proteins are removed, and it also allows for the regulation of cellular functions in response to changing conditions. The balance between protein synthesis and degradation is vital for cellular homeostasis.

- **Protein Synthesis**: The synthesis of proteins begins with the translation of messenger RNA (mRNA) into polypeptide chains by ribosomes. Amino acids are linked together via peptide bonds to form proteins, with the sequence dictated by the mRNA template.
- **Protein Degradation**: Proteins are constantly degraded in cells to prevent the accumulation of dysfunctional or unnecessary proteins. The primary pathways for protein degradation are:
- **Proteasomal degradation**: The proteasome is a large protein complex that degrades proteins tagged with ubiquitin, a small protein that signals that the target protein is to be broken down.
- **Autophagy**: This process involves the lysosomal degradation of entire cellular structures or proteins, often in response to stress or damage.

Protein degradation plays an essential role in regulating cellular processes such as the cell cycle, signaling, and stress responses.

11.2 Amino Acid Synthesis and the Role of Vitamins and Minerals

Amino acids are essential for protein synthesis and serve as precursors for various bioactive molecules. While some amino acids are considered **essential** and must be obtained from the diet, others are **non-essential** and can be synthesized by the body. The synthesis of non-essential amino acids occurs through various pathways, often involving intermediates from metabolic cycles such as the citric acid cycle or glycolysis.

- **Transamination**: This is a key reaction in amino acid synthesis, where an amino group is transferred from one amino acid to a keto acid. For example, **glutamate** can donate an amino group to **alpha-ketoglutarate** to form **glutamine**.
- **Biosynthesis of Specific Amino Acids**: Some amino acids, such as **serine** and **glycine**, can be synthesized from glycolytic intermediates, while others, like **phenylalanine** and **tryptophan**, are derived from precursors in the aromatic amino acid biosynthesis pathway.

Vitamins and minerals are essential cofactors in many of the enzymatic reactions involved in amino acid synthesis. For example:

- **Vitamin B6** (pyridoxine) is a cofactor for **aminotransferases** in the transamination reactions.
- **Folate** (vitamin B9) is involved in the synthesis of purines and pyrimidines, which are important for nucleotide metabolism.
- **Vitamin B12** is critical for the synthesis of **methionine** from **homocysteine**, a reaction that is important for the synthesis of certain amino acids.

11.3 The Role of the Proteasome and Autophagy in Protein Quality Control

Cells need to maintain a delicate balance between synthesizing and degrading proteins to ensure that functional, properly folded proteins are available while dysfunctional proteins are efficiently removed. **Proteasomal degradation** and **autophagy** are central to this protein quality control system.

- **Proteasome Function**: The proteasome is a large, multisubunit complex that recognizes and degrades polyubiquitinated proteins. This process is tightly regulated and plays a crucial role in controlling various cellular processes, including the regulation of the cell cycle, response to stress, and protein quality control.
- **Autophagy**: Autophagy is the process by which cells degrade and recycle damaged or unnecessary cellular components. In autophagy, a portion of the cytoplasm, along with damaged proteins or organelles, is enclosed in a membrane and delivered to the lysosome for degradation. This process is essential for maintaining cellular homeostasis, especially under conditions of nutrient deprivation or cellular stress.

Both proteasomal degradation and autophagy are critical for preventing the accumulation of defective proteins, which can lead to cellular dysfunction and disease.

11.4 Disorders in Protein Metabolism

Disruptions in protein metabolism can result in a variety of metabolic disorders, some of which can have severe consequences for health. Some key disorders related to protein metabolism include:

- **Phenylketonuria (PKU)**: PKU is a genetic disorder caused by a deficiency in the enzyme **phenylalanine hydroxylase**, which converts phenylalanine into tyrosine. As a result, phenylalanine accumulates in the blood and can lead to intellectual disability and neurological damage. Early detection through newborn screening and dietary management is essential to prevent these adverse effects.

- **Maple Syrup Urine Disease (MSUD)**: MSUD is caused by a deficiency in the branched-chain alpha-keto acid dehydrogenase complex, leading to the accumulation of branched-chain amino acids (leucine, isoleucine, and valine) in the blood. This disorder is named for the distinctive sweet odor of the urine in affected individuals. If left untreated, MSUD can lead to neurological damage, developmental delay, and death.

- **Ubiquitin-Proteasome System Disorders**: Mutations in the genes encoding components of the proteasome or ubiquitin ligases can lead to the accumulation of damaged or misfolded proteins in cells. Such disorders are often associated with neurodegenerative diseases, including **Parkinson's disease**, **Alzheimer's disease**, and **Huntington's disease**, where defective protein accumulation is a hallmark feature.

- **Autophagy-Related Disorders**: Defects in autophagy-related genes can lead to diseases such as **Lysosomal Storage Diseases** and **neurodegenerative diseases** like **Parkinson's** and **Alzheimer's**. Impaired autophagy can prevent the clearance of damaged proteins and organelles, contributing to cellular toxicity and organ dysfunction.

11.5 Regulation of Protein Metabolism in Health and Disease

The regulation of protein metabolism is highly dynamic and is influenced by numerous factors, including hormonal signals, nutrient availability, and cellular stress. Key regulators of protein metabolism include:

- **Insulin and Glucagon**: These two hormones play opposing roles in regulating protein metabolism. Insulin promotes protein synthesis by activating protein translation machinery and inhibiting protein degradation, while glucagon, typically released in response to low blood glucose levels, promotes protein catabolism to provide amino acids for gluconeogenesis.
- **Cortisol**: Cortisol, a hormone released in response to stress, stimulates protein catabolism and the breakdown of muscle protein to release amino acids. Chronic high levels of cortisol, as seen in prolonged stress or disease states, can lead to muscle wasting and protein deficiency.
- **Nutrient Sensing Pathways**: The **mTOR (mechanistic target of rapamycin)** pathway plays a central role in regulating protein synthesis in response to nutrient availability. When nutrients like amino acids are abundant, mTOR is activated to promote protein synthesis. Conversely, during nutrient deprivation, mTOR activity is suppressed to conserve energy and protein resources.
- **Autophagy and Proteasomal Pathways**: As mentioned earlier, the autophagy and proteasomal pathways are critical for maintaining protein homeostasis. These pathways are tightly regulated in response to nutrient levels, stress, and cellular damage. For example, during fasting or nutrient deprivation, autophagy is activated to provide the cell with essential amino acids and energy by degrading cellular components.

11.6 Conclusion

Protein metabolism is a complex and highly regulated process that ensures the proper synthesis, turnover, and degradation of proteins within cells. The synthesis of amino acids and the regulation of protein degradation are essential for maintaining cellular function and homeostasis. Disorders in protein metabolism can lead to severe metabolic and neurodegenerative diseases, underscoring the importance of protein quality control mechanisms, such as the proteasome and autophagy. Through understanding these processes, we gain insight into the biochemical foundation of human health and the pathophysiology of various diseases, ultimately contributing to improved therapeutic strategies and treatments.

Chapter 12: Gene Expression and Regulation

Central Dogma of Molecular Biology: DNA → RNA → Protein

At the core of biochemistry lies the principle known as the **central dogma of molecular biology**, which states that genetic information flows from **DNA** to **RNA** and ultimately to **protein**. This flow of information is the foundation for cellular function, guiding everything from metabolism to immune response, and influencing the development and growth of organisms.

DNA (Deoxyribonucleic Acid) serves as the genetic blueprint of life, containing the instructions for building proteins that perform a vast array of tasks in cells. **RNA (Ribonucleic Acid)** acts as the messenger, translating the DNA instructions into a form that can be read by the cell's machinery to assemble proteins. Finally, **proteins** are the workhorses of the cell, carrying out most of the cellular functions and maintaining the structure and integrity of organisms.

Transcription and RNA Processing

Transcription is the first step in the expression of genetic information. During this process, an RNA molecule is synthesized using a DNA template. The enzyme **RNA polymerase** binds to a specific region of the DNA known as the **promoter** and reads the DNA sequence to create a complementary RNA strand. This messenger RNA (mRNA) will carry the genetic instructions from the DNA in the nucleus to the ribosomes in the cytoplasm, where proteins are synthesized.

Before mRNA leaves the nucleus, it undergoes extensive processing, including:

1. **Capping**: The addition of a protective cap to the 5' end of the mRNA.
2. **Splicing**: Removal of **introns** (non-coding regions) and joining of **exons** (coding regions) to create a continuous coding sequence.
3. **Polyadenylation**: Addition of a poly-A tail to the 3' end of the mRNA to protect it from degradation and aid in translation.

This mature mRNA is then transported to the cytoplasm, ready for translation.

Translation and the Role of Ribosomes

Translation is the process by which the mRNA code is read by the **ribosome**, an organelle responsible for protein synthesis. Ribosomes are made up of **ribosomal RNA (rRNA)** and proteins, and they serve as the site where **amino acids** are assembled into proteins according to the instructions encoded in the mRNA.

1. **Initiation**: The small ribosomal subunit binds to the mRNA and locates the **start codon**, signaling the beginning of protein synthesis.
2. **Elongation**: The ribosome moves along the mRNA, reading three nucleotides (a codon) at a time, and adding the corresponding amino acids to the growing polypeptide chain. Each codon specifies a particular amino acid, which is brought to the ribosome by **transfer RNA (tRNA)**.
3. **Termination**: When a **stop codon** is encountered, the ribosome releases the completed polypeptide chain, which will fold into its functional protein structure.

Gene Regulation: Promoters, Enhancers, and Silencers

Gene expression does not occur in a vacuum; it is tightly regulated to ensure that proteins are made only when needed. Regulation of gene expression happens at multiple levels, beginning with the transcriptional control of **gene promoters**.

- **Promoters**: These DNA sequences, located near the start of genes, are binding sites for **transcription factors**—proteins that influence whether RNA polymerase will initiate transcription. Promoters act as switches that turn gene expression on or off.
- **Enhancers**: These regulatory elements can be located far from the gene they control. Enhancers increase the likelihood of transcription by interacting with transcription factors, often facilitating the binding of the transcription machinery to the promoter.
- **Silencers**: Conversely, silencers are DNA sequences that inhibit gene expression by binding to repressor proteins that prevent the transcription machinery from accessing the promoter.

These regulatory elements are crucial in ensuring that genes are expressed at the right time, in the right place, and at the appropriate level. For example, a gene for an enzyme involved in digestion will be expressed in the cells of the digestive system but not in muscle cells.

Epigenetic Modifications and Their Effects on Gene Expression

Beyond the DNA sequence itself, **epigenetic modifications** influence gene expression. These modifications do not change the underlying genetic code but can alter how genes are expressed.

- **DNA Methylation**: The addition of a methyl group (CH_3) to certain cytosine bases in DNA can silence gene expression by making the DNA more tightly packed, preventing the transcription machinery from accessing the gene.
- **Histone Modification**: DNA is wrapped around proteins called histones, and chemical modifications to histones can either relax or condense the DNA, influencing its accessibility for transcription.

Epigenetic changes can be influenced by environmental factors such as diet, stress, and toxins, and they can be passed on to future generations. This field of study is particularly important in understanding how environmental factors can affect gene expression and contribute to diseases such as cancer, obesity, and neurological disorders.

Regulation of Gene Expression in Health and Disease

In healthy cells, gene expression is finely tuned, ensuring that proteins are produced in response to the cell's needs. However, dysregulation of gene expression can lead to diseases. For example:

- **Cancer**: Abnormal gene expression, often caused by mutations in **oncogenes** (genes that promote cell division) or **tumor suppressor genes** (genes that inhibit cell division), can lead to uncontrolled cell proliferation, a hallmark of cancer.
- **Genetic Disorders**: Mutations in genes that encode essential proteins can lead to diseases such as cystic fibrosis, sickle cell anemia, and muscular dystrophy. In some cases, aberrant gene regulation rather than mutations in the gene itself can cause disease. For instance, improper silencing of tumor suppressor genes can contribute to cancer development.

Understanding the mechanisms that regulate gene expression is key to developing targeted therapies for these conditions. **Gene therapy** and **RNA-based treatments** (such as RNA interference or CRISPR-Cas9 gene editing) are emerging fields that aim to correct or alter gene expression in specific cells, offering hope for personalized treatments of various genetic disorders.

Conclusion

Gene expression and regulation are central to life at the molecular level. From the faithful transmission of genetic information in the central dogma to the sophisticated mechanisms that control when, where, and how genes are expressed, this process underpins all cellular functions and ultimately, the functioning of an organism. Disruptions in gene expression can lead to a variety of diseases, but understanding and manipulating these processes holds immense potential for the development of targeted therapies and advances in personalized medicine.

By mastering the concepts of transcription, translation, gene regulation, and epigenetics, we gain deeper insight into the molecular machinery that powers life and opens the door to transformative medical innovations.

Chapter 13: Signal Transduction Pathways

Signal transduction is the process by which cells interpret and respond to external signals, such as hormones, growth factors, or environmental changes. At the molecular level, these signals are translated into biochemical events within the cell, leading to a variety of cellular responses that regulate processes like growth, differentiation, metabolism, and even apoptosis (programmed cell death). Signal transduction pathways play a critical role in maintaining cellular homeostasis and in mediating the effects of external stimuli on cellular behavior.

This chapter explores the mechanisms of cellular signaling, key players in the signaling process, and their implications in health and disease, with a particular focus on the role of signal transduction in growth, differentiation, and cancer.

Overview of Cellular Signaling

At its core, signal transduction involves the conversion of a **chemical signal** (often a ligand) into a functional **biological response**. This process is typically initiated when a signaling molecule (such as a hormone, neurotransmitter, or cytokine) binds to a **receptor** on the surface of the target cell or within the cell. The binding of the ligand causes a conformational change in the receptor, which in turn triggers a cascade of intracellular signaling events.

These signaling pathways involve a series of molecular interactions that transmit the signal from the receptor to various intracellular targets. The signal can result in changes in gene expression, enzyme activity, or the movement of molecules within the cell. Ultimately, the cell responds to the signal by carrying out specific actions, such as cell division, differentiation, or metabolic alterations.

Receptors and Signal Transduction Mechanisms

Signal transduction is initiated by the binding of a signaling molecule (ligand) to its specific receptor. The nature of the receptor largely determines the signaling mechanism and the cellular response. Receptors can be classified into two broad categories based on their location and structure:

Cell Surface Receptors

- **G-protein-coupled receptors (GPCRs)**: These are the largest and most diverse class of receptors, found on the cell membrane. They consist of a single polypeptide chain that spans the membrane seven times. Upon ligand binding, GPCRs activate intracellular **G-proteins**, which in turn activate or inhibit downstream signaling pathways, such as those involving **cAMP, phosphatidylinositol**, or **calcium ions**.
- **Receptor Tyrosine Kinases (RTKs)**: These receptors also reside on the cell membrane but have intrinsic kinase activity. Upon ligand binding (such as growth factors), RTKs undergo **autophosphorylation**, which creates docking sites for downstream signaling proteins. RTKs are often involved in regulating cell growth, survival, and differentiation. A well-known example is the **epidermal growth factor receptor (EGFR)**.
- **Ion Channel Receptors**: These receptors are typically involved in rapid signaling events. They open or close in response to the binding of a ligand, allowing ions such as calcium, sodium, or potassium to flow across the membrane. This leads to changes in the cell's membrane potential, which is critical for processes such as synaptic transmission in neurons.

Intracellular Receptors

These receptors are located within the cell, typically in the cytoplasm or nucleus. They are activated by signaling molecules such as steroid hormones, thyroid hormones, or retinoic acid. Upon ligand binding, the receptor-ligand complex translocates to the nucleus, where it influences gene expression. A key example is the

, which regulate genes involved in metabolism, development, and immune function.

Second Messengers: cAMP, Calcium, IP3

Many signal transduction pathways involve **second messengers**, which amplify the signal and propagate the response within the cell. These molecules are produced in response to receptor activation and act as intracellular mediators.

1. **Cyclic AMP (cAMP)**: cAMP is one of the most common second messengers. It is produced by the enzyme **adenylyl cyclase**, which is activated by GPCRs. cAMP activates **protein kinase A (PKA)**, which in turn phosphorylates target proteins, leading to changes in cellular activity. cAMP is involved in a variety of processes, including the regulation of metabolism, gene expression, and cell growth.

2. **Calcium Ions (Ca^{2+})**: Calcium is a versatile second messenger involved in many signaling pathways. It can be released from intracellular stores, such as the endoplasmic reticulum (ER), or enter the cytoplasm through ion channels. Elevated calcium levels trigger a variety of cellular responses, including muscle contraction, neurotransmitter release, and gene expression.

3. **Inositol Trisphosphate (IP_3) and Diacylglycerol (DAG)**: These second messengers are generated through the activation of phospholipase C (PLC), which cleaves a membrane phospholipid, **PIP_2**, into IP_3 and DAG. IP_3 stimulates the release of calcium from intracellular stores, while DAG activates **protein kinase C (PKC)**, leading to a variety of cellular responses, including changes in metabolism, gene expression, and cell growth.

The Role of Signal Transduction in Cell Growth, Differentiation, and Cancer

Signal transduction is central to regulating processes such as cell growth, differentiation, and survival. Dysregulation of these pathways can lead to diseases, including **cancer**.

Cell Growth and Differentiation

- Growth factors like **epidermal growth factor (EGF)** and **fibroblast growth factor (FGF)** bind to RTKs, triggering intracellular signaling pathways that promote cell proliferation and differentiation. These pathways often activate **MAP kinases**, which regulate the cell cycle and gene expression.
- The **Notch signaling pathway**, a highly conserved pathway, regulates cell fate determination, differentiation, and stem cell maintenance. This pathway is involved in processes such as neural development, blood cell differentiation, and tissue regeneration.

Cancer and Signal Transduction

- **Oncogenes**: In cancer, signal transduction pathways are often disrupted. **Oncogenes** are mutated forms of normal genes (proto-oncogenes) that drive cell proliferation. Many oncogenes encode components of the signaling pathways, such as growth factor receptors or downstream signaling molecules. For instance, mutations in the **EGFR** can lead to uncontrolled cell growth, a hallmark of many cancers.
- **Tumor Suppressors**: Tumor suppressor genes, such as **p53** and **BRCA1**, encode proteins that regulate the cell cycle and promote apoptosis in response to DNA damage. Loss of function of these tumor suppressors can lead to unchecked cell division and cancer development.
- **Warburg Effect**: Cancer cells often exhibit altered metabolism, a phenomenon known as the **Warburg effect**, where they rely heavily on glycolysis for energy production, even in the presence of oxygen. This metabolic shift is linked to the dysregulation of various signaling pathways that influence cellular metabolism.

Cancer Therapy and Targeted Treatments

- Many modern cancer therapies target specific components of signal transduction pathways. For example, **Tyrosine kinase inhibitors (TKIs)**, such as **imatinib** (Gleevec), specifically block the activity of BCR-ABL, a fusion protein formed by the Philadelphia chromosome in chronic myelogenous leukemia (CML).
- Other therapies, such as **monoclonal antibodies** (e.g., trastuzumab, Herceptin), target overexpressed receptors like **HER2** (human epidermal growth factor receptor 2) to inhibit the signaling pathways that promote cancer cell growth.

Conclusion

Signal transduction is a fundamental process by which cells interpret and respond to their environment. By understanding the molecular mechanisms involved in these pathways, we gain insight into how cells regulate crucial processes like growth, differentiation, and metabolism. Furthermore, the dysregulation of signal transduction pathways is central to the development of diseases such as cancer, making these pathways prime targets for therapeutic intervention.

The complexity and diversity of signal transduction mechanisms underline the dynamic nature of cellular regulation, offering opportunities for developing novel treatments for a wide range of conditions. As we continue to explore these pathways, the potential for therapeutic breakthroughs in areas such as cancer, neurobiology, and metabolic disorders remains immense.

Chapter 14: Cellular Communication and Integration

Introduction

Cellular communication is a fundamental aspect of biochemistry that ensures the proper functioning of organisms. From the coordination of simple metabolic processes to the complex behaviors of cells in multicellular organisms, cellular communication governs a vast array of physiological and biochemical functions. In this chapter, we will explore the mechanisms that allow cells to communicate, the integration of metabolic networks, and how systems biology has enhanced our understanding of these processes.

The Role of Hormones in Cellular Communication

Hormones are signaling molecules that travel through the bloodstream to target cells, regulating a wide range of physiological processes. They are primarily produced by endocrine glands, including the pituitary, thyroid, adrenal glands, and pancreas. Hormones can be classified into three main types based on their chemical structure: **peptides**, **steroids**, and **amines**.

- **Peptide hormones** (e.g., insulin, glucagon) are synthesized as larger precursors and require modification to become active. They bind to receptors on the cell surface, initiating a cascade of intracellular signaling events.
- **Steroid hormones** (e.g., cortisol, estrogen) are lipophilic and can pass through cell membranes to bind to intracellular receptors, influencing gene expression directly.
- **Amines** (e.g., thyroid hormone, adrenaline) are derived from amino acids and can also act on cell surface receptors or intracellular targets, depending on their structure.

Hormones regulate critical functions, such as metabolism, growth, and reproduction. For example, **insulin** lowers blood glucose levels by facilitating the uptake of glucose into cells, while **glucagon** works antagonistically to raise blood glucose levels.

Hormonal Regulation of Metabolism

The regulation of metabolism by hormones involves complex feedback loops that allow organisms to respond to changes in their environment and internal conditions. Major metabolic pathways such as glycolysis, the citric acid cycle, and fatty acid metabolism are tightly controlled by hormones.

1. **Insulin and Glucagon**: Insulin promotes an anabolic state by stimulating the synthesis of glycogen, proteins, and fats, while inhibiting the breakdown of these molecules. Glucagon, on the other hand, activates catabolic processes, such as glycogenolysis and gluconeogenesis, to increase blood glucose levels during periods of fasting or stress.
2. **Thyroid Hormones**: Thyroid hormones, such as thyroxine (T4) and triiodothyronine (T3), regulate basal metabolic rate (BMR) by influencing the rate of energy production in cells. An overactive thyroid (hyperthyroidism) results in increased metabolism, while an underactive thyroid (hypothyroidism) leads to decreased metabolic activity.
3. **Corticosteroids**: Produced by the adrenal glands, corticosteroids such as cortisol play a crucial role in the body's response to stress. They stimulate the release of glucose from stored glycogen and promote protein catabolism to provide energy in times of need. Cortisol also has anti-inflammatory effects.

The Nervous System's Role in Biochemical Regulation

In addition to hormones, the nervous system is a major player in cellular communication. Neurons transmit electrical impulses, which are converted into chemical signals that can alter the activity of target cells.

- **Neurotransmitters** are chemical messengers that transmit signals across synapses between neurons or from neurons to muscle cells. Common neurotransmitters include **dopamine, serotonin,** and **acetylcholine**.
- **Neurohormones**, such as **norepinephrine** and **epinephrine**, are secreted by neurons and act on distant target cells, similar to endocrine hormones. These molecules are especially important during the fight-or-flight response, where they rapidly prepare the body for action.
- **The autonomic nervous system** (ANS) regulates involuntary functions like heart rate, digestion, and respiration through sympathetic and parasympathetic pathways. The sympathetic nervous system triggers the release of catecholamines like epinephrine, which increase heart rate and energy availability, while the parasympathetic nervous system promotes rest and digestion.

Cross-Talk Between Metabolic Pathways

Metabolic pathways do not function in isolation. Instead, they are interconnected, allowing the cell to respond to various signals and maintain homeostasis. Cross-talk between pathways ensures efficient energy usage and adaptation to changing conditions. Some key examples of cross-talk include:

- **Glycolysis and Gluconeogenesis**: These two pathways are reciprocally regulated. When energy is abundant, glycolysis is favored, while during fasting or energy depletion, gluconeogenesis takes precedence to generate glucose from non-carbohydrate sources.
- **The Urea Cycle and Amino Acid Metabolism**: The urea cycle is tightly linked with amino acid metabolism. When proteins are broken down, ammonia is produced as a byproduct. The urea cycle detoxifies ammonia by converting it to urea, which is excreted by the kidneys.
- **Lipid and Carbohydrate Metabolism**: The breakdown of fats (lipolysis) and carbohydrates (glycogenolysis) provides intermediates that feed into common metabolic pathways, such as the citric acid cycle. This integration ensures that energy needs are met regardless of the macronutrient source.

Systems Biology and the Integration of Metabolic Networks

Systems biology is an interdisciplinary field that uses computational and mathematical models to analyze the interactions and functions of biological systems. By studying the integration of cellular networks, researchers can gain a deeper understanding of cellular communication and its role in health and disease.

In the context of metabolism, systems biology allows for the modeling of complex biochemical pathways and the prediction of how changes in one component (e.g., enzyme activity, gene expression) might affect the entire system. This approach has proven valuable in understanding diseases such as cancer, diabetes, and neurodegenerative disorders, where disruptions in cellular communication and metabolism play key roles.

Key tools in systems biology include:

- **Omics technologies**: These include genomics, proteomics, metabolomics, and transcriptomics, which provide detailed information about the molecules present in a cell and how they change in response to various stimuli.
- **Bioinformatics**: The use of computational tools to analyze large datasets generated by omics technologies enables researchers to identify patterns and relationships between molecular components.
- **Pathway analysis**: Systems biology often involves the reconstruction of metabolic and signaling pathways, which helps researchers visualize how cellular processes are interconnected and how perturbations in these pathways lead to disease.

Conclusion

Cellular communication and integration are essential for maintaining the proper functioning of biological systems. Hormones, the nervous system, and cross-talk between metabolic pathways work together to ensure the efficient regulation of metabolism, growth, and adaptation to environmental changes. Systems biology is providing new insights into these processes, offering a holistic understanding of how cells interact and function within larger biological networks. By continuing to unravel these intricate molecular interactions, biochemists can pave the way for innovations in medicine, biotechnology, and health.

Chapter 16: Biochemical Techniques and Technologies
Introduction

The field of biochemistry has advanced significantly, not only through theoretical understanding but also by the development of experimental techniques that allow scientists to study biological systems at the molecular level. Modern biochemical techniques provide the tools necessary to analyze the structure, function, and interactions of molecules within cells, tissues, and organisms. In this chapter, we will explore some of the key laboratory techniques and technologies used in biochemistry today, including chromatography, electrophoresis, mass spectrometry, protein analysis, DNA sequencing, and the role of bioinformatics.

Chromatography

Chromatography is a powerful technique for separating mixtures of substances based on differences in their physical or chemical properties, such as polarity, size, or charge. It is widely used in biochemistry to purify proteins, nucleic acids, and small molecules. There are several types of chromatography, each suited to different applications:

- **Thin-layer chromatography (TLC)**: A simple technique where the sample is applied to a solid adsorbent (usually a silica gel) on a flat surface. The sample is then separated based on its interaction with a solvent that moves up the plate by capillary action.
- **Column chromatography**: A more advanced form of chromatography, where a mixture is passed through a column packed with an adsorbent material, allowing separation based on different retention times. Variations include **ion-exchange chromatography**, which separates proteins based on their charge, and **affinity chromatography**, which separates molecules based on specific binding interactions.
- **High-performance liquid chromatography (HPLC)**: A highly refined and automated form of column chromatography, used to separate, identify, and quantify compounds in complex mixtures. HPLC is commonly used for protein purification, enzyme assays, and analysis of metabolites.

Electrophoresis

Electrophoresis is a technique used to separate charged molecules (typically proteins or nucleic acids) based on their size and charge by applying an electric field. The most common types of electrophoresis are:

- **Agarose gel electrophoresis**: Primarily used for the separation of nucleic acids such as DNA and RNA. When an electric field is applied, DNA molecules migrate towards the positive electrode due to their negative charge, with smaller molecules traveling faster than larger ones. The gel can be stained with a dye, such as ethidium bromide, to visualize the separated bands under UV light.
- **Polymerase chain reaction (PCR)**: Often used in conjunction with electrophoresis to amplify DNA segments before analysis. PCR generates millions of copies of a specific DNA sequence, which can then be separated using agarose gel electrophoresis.
- **SDS-PAGE (Sodium dodecyl sulfate polyacrylamide gel electrophoresis)**: A technique used to separate proteins by size. Proteins are first treated with SDS, which denatures them and gives them a uniform negative charge. The proteins are then loaded into a gel and subjected to an electric field. The smaller proteins move through the gel faster than larger ones.

Mass Spectrometry

Mass spectrometry (MS) is an analytical technique used to measure the mass-to-charge ratio of ions. In biochemistry, it is commonly used to identify and characterize proteins, peptides, metabolites, and other small molecules. Mass spectrometry can provide detailed information about the molecular weight, structure, and composition of a sample.

The typical mass spectrometry workflow includes the following steps:

1. **Ionization**: The sample is ionized (e.g., by **electrospray ionization** or **matrix-assisted laser desorption/ionization**), producing charged particles.
2. **Analysis**: The ions are separated based on their mass-to-charge ratio in a mass analyzer (e.g., quadrupole, time-of-flight (TOF)).
3. **Detection**: The separated ions are detected and quantified based on their intensity.

Mass spectrometry is an essential tool in proteomics (the study of proteins), where it can be used to identify proteins, analyze post-translational modifications, and study protein-protein interactions. It is also employed in metabolomics to analyze metabolic profiles and detect biomarkers.

Techniques for Protein Analysis

Proteins are the central components of cellular machinery, and understanding their structure and function is a key goal of biochemistry. Several techniques are used to analyze proteins:

- **Western blotting (Immunoblotting)**: A technique used to detect specific proteins in a sample. The process involves separating proteins by SDS-PAGE, transferring them to a membrane, and then using antibodies to bind and detect the target protein. The presence and abundance of the protein are visualized by chemiluminescence or colorimetric methods.
- **Enzyme-linked immunosorbent assay (ELISA)**: A quantitative technique that detects the presence of proteins or other antigens in a sample. It uses an antibody linked to an enzyme that produces a detectable signal, typically a color change, when bound to its target.
- **Circular dichroism (CD) spectroscopy**: A technique that analyzes the secondary structure of proteins by measuring the differential absorption of left- and right-handed circularly polarized light. It is used to study protein folding, stability, and conformational changes.

DNA Sequencing and PCR

The sequencing of DNA has revolutionized biochemistry, enabling the identification of genetic sequences and the study of gene expression. The two major DNA sequencing techniques are:

- **Sanger sequencing**: The first widely used method for DNA sequencing, based on chain-termination. In this method, DNA is amplified and then sequenced by adding labeled dideoxynucleotides (which terminate DNA elongation), producing fragments of different lengths that are subsequently read by a detector.
- **Next-generation sequencing (NGS)**: A more advanced technology that allows for high-throughput sequencing of large DNA fragments. NGS can sequence entire genomes rapidly and is used for applications such as whole-genome sequencing, transcriptomics, and metagenomics.

Polymerase chain reaction (PCR) is another crucial technique in modern biochemistry. PCR allows the amplification of specific DNA sequences, making it possible to obtain large amounts of DNA for further analysis. PCR is essential in diagnostic applications, such as detecting genetic mutations and pathogens, and for cloning DNA fragments.

The Role of Bioinformatics in Modern Biochemistry

Bioinformatics is the field that combines biology, computer science, and mathematics to analyze and interpret biological data. The increasing availability of large-scale biological data sets, such as genome sequences and protein structures, has made bioinformatics indispensable in modern biochemistry.

- **Sequence alignment and comparison**: Bioinformatics tools can align DNA, RNA, or protein sequences to identify similarities and differences. This is useful in gene identification, evolutionary studies, and mutation analysis.
- **Structural bioinformatics**: This involves predicting and analyzing the three-dimensional structures of proteins and nucleic acids. Tools like **protein docking** and **molecular dynamics simulations** allow researchers to study how biomolecules interact, which is crucial for drug design and protein engineering.
- **Systems biology and network analysis**: Bioinformatics plays a central role in systems biology by integrating data from various "omics" fields (genomics, proteomics, metabolomics) to model biological processes and metabolic networks. These models help scientists understand cellular functions, identify disease mechanisms, and discover new therapeutic targets.

Conclusion

Biochemical techniques and technologies are essential for exploring the molecular foundations of life. Methods like chromatography, electrophoresis, mass spectrometry, and DNA sequencing have enabled tremendous advances in the understanding of biochemistry. In addition, bioinformatics has become a cornerstone of modern biochemical research, facilitating the analysis of complex data and enabling new discoveries in fields such as drug development, systems biology, and genomics. As technology continues to evolve, these tools will be critical in unraveling the remaining mysteries of life at the molecular level, contributing to advances in medicine, biotechnology, and personalized healthcare.

Chapter 17: The Role of Biochemistry in Drug Discovery
Introduction

The process of drug discovery is complex, interdisciplinary, and requires a deep understanding of the biological systems targeted by therapeutic agents. Biochemistry plays a pivotal role in identifying, designing, and developing drugs that can modulate biological pathways to treat diseases. From the initial identification of disease targets to the final stages of clinical trials, biochemistry is integral in creating drugs that are safe, effective, and tailored to the molecular mechanisms underlying various health conditions. In this chapter, we will explore how biochemistry shapes the drug discovery process, focusing on target identification, structure-based drug design, biochemical assays, and case studies of successful drug development.

Target Identification and Validation

The first step in drug discovery is identifying and validating a molecular target—typically a protein, nucleic acid, or small molecule involved in the disease process. This step requires an understanding of the biochemical pathways involved in the disease and how altering a specific biomolecule can have therapeutic effects.

- **Target Identification**: Target identification begins with understanding the biological mechanisms underlying a disease. In cancer, for example, the discovery of mutations in oncogenes or tumor suppressor genes can reveal potential drug targets. For infectious diseases, the structure of viral or bacterial proteins is often the focus of target identification. Advanced techniques like **genomics**, **proteomics**, and **metabolomics** provide comprehensive data on the molecular players involved in diseases, making target identification more precise.
- **Target Validation**: After identifying a potential target, it must be validated to ensure that modulating it will result in a therapeutic effect. This often involves genetic knockdowns, overexpression studies, or the use of small molecule inhibitors to confirm that the target is involved in the disease. For example, knocking down the expression of a target protein in a cell culture or animal model should lead to a reduction in disease symptoms or progression.

Structure-Based Drug Design

Once a target has been identified and validated, the next step is to design molecules that can interact with it in a specific, controlled manner. **Structure-based drug design** (SBDD) uses the 3D structure of a target protein to design drugs that bind selectively and effectively.

- **X-ray Crystallography and NMR Spectroscopy**: To facilitate structure-based drug design, the 3D structure of a protein or enzyme must be determined. Techniques like **X-ray crystallography** and **nuclear magnetic resonance (NMR) spectroscopy** allow researchers to visualize the precise shape and active site of a protein. This information is critical in designing molecules that will bind to the target with high affinity and specificity.
- **Computational Drug Design**: Once the structure of a target is available, computational tools are used to predict how small molecules can interact with the protein. **Molecular docking** is a common technique used to simulate the binding of potential drug candidates to the target protein's active site. The goal is to find small molecules that fit well into the binding pocket and block or enhance the target protein's function.
- **Virtual Screening**: Computational methods allow for **virtual screening** of large libraries of compounds to identify potential drug candidates that can bind to the target. This method accelerates the drug discovery process by narrowing down the vast number of possible compounds to a manageable set of promising candidates.

Biochemical Assays in Drug Screening

Once potential drug candidates have been identified, biochemical assays are used to screen these compounds for their ability to interact with the target and produce the desired effect. These assays are critical for evaluating the efficacy, potency, and specificity of a drug before further development.

- **Enzyme Inhibition Assays**: In cases where the target is an enzyme, enzyme inhibition assays are used to measure the ability of a compound to reduce the enzyme's activity. This is often the first step in drug screening, and the most promising inhibitors can then be subjected to more detailed testing.

- **Cell-Based Assays**: For targets located within cells, cell-based assays are used to evaluate the compound's ability to affect cellular processes. These assays allow researchers to study the drug's effect in the context of the whole cell, including its impact on cell growth, signaling pathways, and gene expression.

- **High-Throughput Screening (HTS)**: HTS is an automated technique that allows researchers to rapidly test large numbers of compounds for biological activity. This method can screen thousands of compounds in a short period, accelerating the identification of potential drug candidates. HTS uses robotic systems to mix compounds with biological targets, and the results are analyzed for activity.

- **ADMET Testing**: After initial screening, the drug candidates are subjected to **ADMET (Absorption, Distribution, Metabolism, Excretion, and Toxicity) testing**. This phase assesses how the drug behaves in the body and helps predict its safety and efficacy in humans. ADMET testing is essential for identifying compounds that will not cause harmful side effects when administered.

Case Studies of Biochemically-Targeted Drugs

Biochemically targeted drugs have led to some of the most important breakthroughs in modern medicine. These drugs work by directly interacting with specific molecules that drive disease processes. Here are a few notable examples:

- **Imatinib (Gleevec)**: Imatinib is a tyrosine kinase inhibitor that targets the BCR-ABL fusion protein, which is found in chronic myelogenous leukemia (CML). This fusion protein is a result of a genetic mutation and is a key driver of CML. Imatinib was developed through the understanding of the biochemical mechanisms behind the disease and has revolutionized the treatment of CML.
- **Trastuzumab (Herceptin)**: Trastuzumab is a monoclonal antibody that targets the HER2 receptor, which is overexpressed in some breast cancers. By binding to HER2, trastuzumab blocks the receptor's signaling and prevents the growth of cancer cells. This targeted therapy has significantly improved the prognosis for patients with HER2-positive breast cancer.
- **Statins**: Statins are a class of drugs used to lower cholesterol levels by inhibiting the enzyme **HMG-CoA reductase**, which plays a key role in cholesterol biosynthesis. The development of statins was based on an understanding of lipid metabolism and the biochemical regulation of cholesterol production in the liver.
- **Cisplatin**: Cisplatin is a chemotherapy drug that forms DNA cross-links, disrupting DNA replication and transcription. It is effective against a wide range of cancers, including testicular, ovarian, and lung cancers. The development of cisplatin was rooted in the understanding of DNA structure and the biochemical basis of cellular replication.

Challenges and Future Directions

While biochemistry has made significant advances in drug discovery, several challenges remain:

- **Drug Resistance**: Over time, cancer cells, viruses, and bacteria can evolve resistance to drugs, rendering treatments less effective. Biochemical research is focusing on identifying new drug targets and developing combination therapies to overcome resistance.
- **Personalized Medicine**: With the advent of genomics, there is increasing interest in personalized medicine, which tailors drug treatments based on an individual's genetic profile. Biochemistry plays a central role in understanding how genetic variations affect drug metabolism, efficacy, and toxicity.
- **Biologics and Gene Therapy**: The rise of biologic drugs, including monoclonal antibodies and gene therapies, presents new challenges in drug discovery. These therapies target specific molecules or correct genetic defects, and biochemistry is essential for developing and optimizing these treatments.
- **Artificial Intelligence in Drug Discovery**: Advances in artificial intelligence (AI) and machine learning are transforming the drug discovery process. AI algorithms can predict how molecules will interact with biological targets, analyze large datasets, and accelerate the identification of promising drug candidates.

Conclusion

Biochemistry is at the heart of drug discovery, driving the development of targeted therapies that are transforming medicine. From target identification to structure-based drug design, biochemical assays, and the case studies of successful drugs, biochemistry plays an indispensable role in creating new treatments for diseases. The future of drug discovery is bright, with ongoing innovations in personalized medicine, biologics, and AI that promise to accelerate the development of novel therapeutic agents. As our understanding of biochemistry deepens, so too will our ability to unlock the full potential of drugs in the fight against disease.

Chapter 18: Metabolic Diseases and Disorders
Introduction

Metabolism is the set of chemical reactions that occur within living organisms to maintain life. These reactions are crucial for converting nutrients into energy, synthesizing molecules needed for growth and repair, and maintaining homeostasis. However, when these metabolic processes are disrupted due to genetic mutations, environmental factors, or other health issues, metabolic diseases and disorders can arise. This chapter explores inherited metabolic diseases, the consequences of genetic mutations on metabolism, current diagnostic and treatment strategies, advances in gene therapy, and the ethical considerations surrounding the treatment of metabolic diseases.

Overview of Inherited Metabolic Diseases

Inherited metabolic diseases (IMDs) are a group of disorders caused by defects in the enzymes or proteins that regulate metabolic pathways. These genetic mutations often result in the accumulation of toxic metabolites or a deficiency in essential compounds, leading to severe health complications. Many of these diseases are rare, but they can have profound impacts on affected individuals and their families.

- **Phenylketonuria (PKU)**: PKU is one of the most well-known inherited metabolic diseases. It occurs due to a mutation in the **PAH gene**, which encodes the enzyme **phenylalanine hydroxylase**. This enzyme is responsible for converting the amino acid phenylalanine into tyrosine. In PKU, the accumulation of phenylalanine in the blood can lead to brain damage, intellectual disability, and other neurological issues if not treated early. Treatment involves a strict low-phenylalanine diet to prevent toxic buildup.
- **Tay-Sachs Disease**: Tay-Sachs is a neurodegenerative disorder caused by a deficiency of the enzyme **hexosaminidase A**, which is needed to break down certain lipids in the nervous system. As a result, lipids accumulate in the brain, leading to progressive neurological damage and early death. It is most common in individuals of Ashkenazi Jewish descent.
- **Gaucher's Disease**: This genetic disorder occurs due to a deficiency in the enzyme **glucocerebrosidase**, which is responsible for breaking down a lipid called **glucocerebroside**. When this enzyme is deficient, glucocerebroside accumulates in the liver, spleen, and bone marrow, causing organ enlargement, bone pain, and other symptoms. Gaucher's disease is treatable with enzyme replacement therapy (ERT).

- **Maple Syrup Urine Disease (MSUD)**: MSUD is caused by defects in the enzymes responsible for breaking down branched-chain amino acids (BCAAs)—leucine, isoleucine, and valine. The buildup of these amino acids leads to neurological damage, with symptoms that often appear in infancy. Immediate dietary restriction of BCAAs can prevent severe outcomes.

- **Mitochondrial Diseases**: Mitochondria are responsible for energy production in cells, and defects in mitochondrial genes can lead to a variety of diseases. These include **Leber's hereditary optic neuropathy** (LHON), a condition that causes vision loss, and **Kearns-Sayre syndrome**, a multisystem disorder that affects the heart, muscles, and other organs.

These examples represent just a few of the hundreds of known inherited metabolic diseases. Each disease involves a specific biochemical pathway that is disrupted, often leading to severe clinical manifestations if not diagnosed and treated promptly.

Metabolic Consequences of Genetic Mutations

Metabolic diseases are often the result of mutations that impair enzymes involved in critical biochemical pathways. These mutations can lead to the accumulation of intermediates that are toxic to cells or organs, or they can result in a deficiency of essential metabolites required for normal cellular function.

- **Enzyme Deficiency**: Many metabolic diseases, such as PKU and Gaucher's disease, are caused by a deficiency in a single enzyme. Enzyme deficiencies lead to the incomplete metabolism of substrates, causing the accumulation of toxic metabolites or the inability to produce necessary products. For example, in PKU, the lack of phenylalanine hydroxylase results in the buildup of phenylalanine, which interferes with brain development.

- **Accumulation of Metabolites**: In diseases like Tay-Sachs and Gaucher's disease, the accumulation of undigested metabolites, such as lipids or sugars, can overwhelm the capacity of cells to remove or store them, leading to cellular dysfunction. For instance, in Tay-Sachs, the accumulation of gangliosides in nerve cells disrupts normal brain function, resulting in neurological deterioration.

- **Impaired Energy Production**: Mitochondrial disorders, caused by mutations in mitochondrial DNA or nuclear genes affecting mitochondrial function, impair cellular energy production. These diseases often affect high-energy tissues, such as muscles and nerves, leading to symptoms like muscle weakness, fatigue, and neurological deficits.

Diagnosis and Treatment Strategies for Metabolic Disorders

The diagnosis of metabolic diseases often begins with a comprehensive evaluation of clinical symptoms, followed by biochemical and genetic testing. Advances in genomics and diagnostic technologies have enabled early detection, particularly for diseases that affect infants and children.

- **Newborn Screening**: Many countries now perform routine newborn screening for metabolic diseases, which can detect conditions such as PKU, MSUD, and galactosemia within the first few days of life. Early diagnosis allows for prompt intervention, which can prevent or mitigate the long-term effects of the disease.
- **Genetic Testing**: Genetic testing plays a critical role in diagnosing inherited metabolic diseases. Identifying the specific mutation responsible for the disease can confirm the diagnosis and help guide treatment decisions. Whole exome sequencing (WES) and whole genome sequencing (WGS) are powerful tools in identifying rare and complex metabolic disorders.
- **Biochemical Testing**: Biochemical tests, such as measuring the levels of specific metabolites or enzymes in the blood, urine, or tissues, can provide important diagnostic clues. For example, high levels of phenylalanine in the blood are indicative of PKU, while the presence of glucocerebroside in urine can help diagnose Gaucher's disease.
- **Dietary and Enzyme Replacement Therapies**: Many inherited metabolic diseases can be managed through dietary interventions. In diseases like PKU, a strict low-phenylalanine diet can prevent the toxic buildup of phenylalanine. In Gaucher's disease, enzyme replacement therapy (ERT) can provide the missing glucocerebrosidase enzyme, alleviating symptoms and improving organ function. Gene therapy, which aims to correct the genetic mutations underlying these diseases, is also being explored as a potential treatment.

- **Liver Transplants and Organ Replacement**: In some cases, liver transplants are considered when the liver is significantly damaged due to metabolic buildup or when enzyme replacement therapy is not effective. Other organ transplants, such as bone marrow transplants, have been used to treat diseases like Gaucher's disease.

Advances in Gene Therapy and Personalized Medicine

Gene therapy holds significant promise for treating genetic metabolic diseases by directly correcting the genetic mutations that cause these disorders. The goal of gene therapy is to replace the defective gene with a functional copy, either by delivering a healthy gene to the patient's cells or by editing the patient's existing DNA.

- **Gene Editing**: Techniques like **CRISPR-Cas9** offer a revolutionary approach to gene therapy by allowing precise modifications to the DNA sequence. In metabolic diseases, gene editing could potentially correct mutations at the molecular level, restoring normal enzyme function and preventing the disease.

- **Viral Vectors**: Viral vectors, such as adeno-associated viruses (AAVs), are commonly used to deliver therapeutic genes into patient cells. Research is ongoing to optimize the use of these vectors to improve their efficiency, safety, and long-term effects in treating metabolic disorders.

- **Personalized Medicine**: Advances in genomics have made personalized medicine a reality. By analyzing an individual's genetic makeup, physicians can tailor treatments based on the specific mutations and metabolic needs of the patient. This approach is particularly useful for rare and complex disorders, where standard treatments may not be effective.

Ethical Considerations in the Treatment of Metabolic Diseases

The treatment of metabolic diseases raises several ethical questions, particularly regarding gene therapy and the use of advanced genetic technologies.

- **Access to Treatment**: Gene therapy and enzyme replacement therapy are often expensive, raising concerns about access to treatment, particularly in low-income regions. Ensuring equitable access to life-saving treatments is an ongoing challenge in the field of metabolic disease management.
- **Gene Editing and Germline Modifications**: The possibility of editing the human germline (the DNA of embryos or reproductive cells) raises ethical concerns about the long-term effects and unintended consequences of altering the human genome. Should parents be allowed to choose traits for their children, and if so, where should we draw the line?
- **Informed Consent**: Patients and their families must be fully informed about the potential risks and benefits of emerging therapies. This includes understanding the long-term effects of gene therapy and the possibility of side effects, including immune reactions to viral vectors.

Conclusion

Metabolic diseases and disorders highlight the delicate balance required to maintain proper biochemical function within the body. While many of these conditions are rare, they underscore the importance of understanding the biochemical pathways that regulate metabolism. Advances in diagnostic technologies, gene therapy, and personalized medicine offer hope for improving the lives of those affected by these diseases. As research continues, it is essential to address the ethical, social, and economic implications of these breakthroughs, ensuring that the benefits of medical advancements are accessible and ethically sound for all.

Chapter 19: Biochemistry of Cancer

Cancer is a complex disease characterized by uncontrolled cell growth and spread to other parts of the body. From a biochemical perspective, cancer can be understood as a series of molecular changes in cells that disrupt normal regulatory mechanisms. These changes can affect various biochemical pathways, including cell cycle control, apoptosis (programmed cell death), DNA repair, and metabolism. In this chapter, we will explore the biochemical underpinnings of cancer, focusing on the role of oncogenes and tumor suppressor genes, the metabolic alterations that support cancer growth, and the latest advancements in targeted cancer therapies.

The Biochemical Basis of Cancer Development

Cancer arises from mutations in the DNA of a cell, which can be caused by environmental factors such as radiation, carcinogens, and viral infections, or by inherited genetic predispositions. These mutations lead to the activation of oncogenes or the inactivation of tumor suppressor genes, resulting in abnormal cell behavior.

1. **Oncogenes** are genes that, when mutated or overexpressed, promote cancer development. They are usually derived from normal genes called proto-oncogenes, which encode proteins involved in cell growth and differentiation. Oncogenes drive uncontrolled cell proliferation by activating signaling pathways that stimulate cell division. For example, the **ras gene** is one of the most well-known oncogenes, and its mutation leads to constant activation of cell growth signaling.

2. **Tumor suppressor genes** normally function to prevent uncontrolled cell growth by regulating the cell cycle, promoting DNA repair, or initiating apoptosis in damaged cells. Mutations in these genes lead to a loss of control over cell division and survival. The **p53 gene**, often called the "guardian of the genome," is one of the most important tumor suppressor genes. When mutated, p53 loses its ability to activate DNA repair or trigger cell death, allowing damaged cells to survive and proliferate.

3. **DNA repair genes**: Cancer cells often exhibit defects in the machinery responsible for repairing DNA damage. The accumulation of genetic mutations leads to genomic instability, which fuels cancer progression. For example, defects in **BRCA1** and **BRCA2**, DNA repair genes, are linked to an increased risk of breast and ovarian cancers.

The Role of Metabolism in Cancer: The Warburg Effect

One of the hallmarks of cancer cells is their altered metabolism. Cancer cells rely heavily on glycolysis, even in the presence of oxygen, a phenomenon known as the **Warburg effect**. This shift in metabolic pathways helps cancer cells meet the high energy demands of rapid growth and proliferation.

1. **Glycolysis and lactate production**: In normal cells, glucose is converted into pyruvate, which is then fed into the mitochondria for oxidative phosphorylation and ATP production. In cancer cells, however, glucose is primarily converted into lactate, even in the presence of oxygen. This process, known as aerobic glycolysis, is less efficient in terms of ATP production but allows cancer cells to divert metabolic intermediates into biosynthetic pathways necessary for cell growth, such as nucleotide and amino acid synthesis.

2. **Mitochondrial dysfunction**: While cancer cells may rely on glycolysis for energy, mitochondria are still essential for other cellular functions, including the regulation of cell death (apoptosis). However, mutations in mitochondrial DNA or changes in mitochondrial function can contribute to the metabolic reprogramming observed in cancer cells.

3. **Fatty acid metabolism**: In addition to glucose, many cancer cells also upregulate fatty acid metabolism. Fatty acids serve as building blocks for the synthesis of cell membranes and signaling molecules. Increased fatty acid synthesis and oxidation provide the necessary lipids for rapidly dividing cells.

4. **Amino acid metabolism**: Cancer cells often have altered amino acid metabolism to support their biosynthetic needs. For example, the amino acid **glutamine** is frequently upregulated in cancer cells, providing an alternative energy source and supporting the synthesis of nucleotides and other metabolites.

Understanding the altered metabolism of cancer cells has led to the development of new therapeutic strategies that target these metabolic changes. Some of the most promising approaches include:

1. **Inhibiting glycolysis**: Drugs that inhibit key enzymes in the glycolytic pathway, such as **hexokinase** or **pyruvate kinase M2**, are being explored as potential cancer treatments. By disrupting the Warburg effect, these drugs can deprive cancer cells of the energy and biosynthetic precursors they need to proliferate.
2. **Targeting mitochondrial function**: Some cancer therapies focus on restoring normal mitochondrial function or inducing mitochondrial-dependent cell death (apoptosis) in cancer cells. For example, compounds that disrupt mitochondrial membrane potential or activate mitochondrial apoptotic pathways are being investigated.
3. **Inhibiting fatty acid synthesis**: Given the importance of fatty acid metabolism in cancer, inhibitors of **fatty acid synthase** (FASN) and other enzymes involved in lipid metabolism are being developed as potential cancer treatments. These drugs aim to starve cancer cells of the lipids necessary for their growth.
4. **Glutamine metabolism inhibitors**: Since glutamine is critical for cancer cell growth, drugs that target **glutaminase**, the enzyme that converts glutamine to glutamate, are being explored as potential cancer therapies.

Cancer Biomarkers and Diagnostic Tools

Early detection of cancer is crucial for improving outcomes. Biochemical markers, or **biomarkers**, are substances produced by cancer cells or by the body in response to cancer. These biomarkers can be measured in blood, urine, or tissue samples and provide valuable information about cancer type, stage, and prognosis.

1. **Oncogene activation**: The presence of specific oncogene mutations or overexpression can be detected through PCR-based techniques or next-generation sequencing (NGS). For example, the **HER2** oncogene is overexpressed in some breast cancers, and its detection can guide treatment decisions.
2. **Tumor suppressor gene inactivation**: Mutations in tumor suppressor genes such as **p53** can be identified using molecular techniques, providing insight into cancer progression and prognosis.
3. **Metabolic biomarkers**: Cancer cells often produce unique metabolic byproducts, such as increased lactate or altered lipid profiles. Mass spectrometry and other analytical techniques are used to identify these metabolic biomarkers, which can aid in early diagnosis and monitoring of treatment efficacy.
4. **Immunohistochemistry (IHC)**: This technique is commonly used to detect the expression of specific proteins, such as **Ki-67** (a marker of cell proliferation) or **p53** (a tumor suppressor), in tissue samples. These markers help assess the aggressiveness of the tumor and predict patient outcomes.

Conclusion

Cancer is a multifaceted disease with profound biochemical underpinnings. Understanding the molecular mechanisms driving cancer—such as the roles of oncogenes, tumor suppressor genes, and altered metabolism—has opened the door to new diagnostic tools and targeted therapies. By focusing on the metabolic vulnerabilities of cancer cells, researchers are developing novel treatments that could revolutionize cancer care. With advances in genomics, metabolic profiling, and personalized medicine, the future of cancer biochemistry holds promise for more effective therapies and better outcomes for patients worldwide.

Chapter 20: Aging and Biochemistry

Aging is an inevitable biological process, but its molecular underpinnings are becoming clearer through the lens of biochemistry. At its core, aging involves a series of biochemical changes that accumulate over time, contributing to cellular dysfunction, tissue degradation, and the eventual decline in physiological function. In this chapter, we will explore the biochemical processes that drive aging, focusing on oxidative stress, telomere shortening, cellular senescence, and the role of longevity genes. We will also discuss the biochemistry of age-related diseases and current interventions aimed at slowing or reversing the aging process.

The Biochemical Processes Involved in Aging

Aging is a complex and multifaceted phenomenon, influenced by genetic, environmental, and lifestyle factors. Biochemically, aging can be attributed to a variety of molecular processes that impair cellular function and increase vulnerability to disease.

1. **Oxidative Stress**

   One of the central biochemical processes implicated in aging is **oxidative stress**, which arises from the accumulation of reactive oxygen species (ROS) in cells. ROS are highly reactive molecules, such as superoxide anions and hydrogen peroxide, that are generated as byproducts of normal cellular metabolism, particularly during oxidative phosphorylation in mitochondria.

   Under normal conditions, ROS are neutralized by antioxidant systems, including enzymes like **superoxide dismutase** (SOD), **catalase**, and **glutathione peroxidase**. However, with age, the efficiency of these antioxidant defenses declines, leading to the accumulation of ROS. These molecules damage cellular components, including proteins, lipids, and DNA, resulting in cellular dysfunction and contributing to the aging process.

   Oxidative stress has been linked to a variety of age-related diseases, such as **cardiovascular diseases, neurodegenerative diseases**, and **cancer**. The accumulation of oxidative damage in tissues compromises their ability to function properly, leading to the gradual decline seen with aging.

2. Telomere Shortening

Telomeres are repetitive DNA sequences located at the ends of chromosomes, protecting them from degradation and preventing chromosome fusion. Every time a cell divides, the telomeres shorten slightly. When telomeres become critically short, cells can no longer divide and enter a state known as **senescence**. Telomere shortening has been implicated as a key factor in aging and age-related diseases. The enzyme **telomerase** can extend telomeres by adding repeats to the chromosome ends, but it is not active in most somatic cells. Some tissues, such as stem cells and germ cells, have higher levels of telomerase, which allows them to maintain telomere length and avoid senescence. However, in most cells, telomere shortening is an irreversible process that limits the number of times a cell can divide.

Studies have shown that **telomere length** is associated with lifespan in many species. Shorter telomeres are associated with an increased risk of age-related diseases, while longer telomeres are linked to improved health and longevity.

3. **Cellular Senescence**

   Cellular senescence refers to the irreversible growth arrest of cells that have been damaged or have reached the limit of their replicative potential (due to telomere shortening). Senescent cells remain metabolically active but no longer divide. These cells accumulate over time and secrete pro-inflammatory factors, contributing to chronic inflammation, a hallmark of aging.

   Senescent cells are thought to play a key role in the aging process by contributing to tissue dysfunction and the development of age-related diseases. For example, in aging tissues, the accumulation of senescent cells can impair wound healing, tissue repair, and immune function.

4. **Mitochondrial Dysfunction**

   Mitochondria are the powerhouses of the cell, responsible for producing ATP through oxidative phosphorylation. As cells age, mitochondrial function declines due to oxidative damage to mitochondrial DNA (mtDNA) and proteins. This mitochondrial dysfunction leads to reduced ATP production and increased ROS generation, further exacerbating oxidative stress.

   The decline in mitochondrial function is thought to contribute to various age-related diseases, including neurodegenerative disorders such as **Alzheimer's disease** and **Parkinson's disease**. The accumulation of damaged mitochondria may also play a role in the aging of muscle tissue, leading to frailty and sarcopenia.

In recent years, **sirtuins**—a family of NAD+-dependent deacetylases—have emerged as key regulators of aging and longevity. Sirtuins regulate various biochemical pathways, including **DNA repair, metabolism, and inflammation**, all of which are central to the aging process.

1. **Sirtuin 1 (SIRT1)**:

   SIRT1 is one of the most studied sirtuins and has been shown to extend lifespan in various organisms, including yeast, worms, and mice. SIRT1 regulates several key processes involved in aging, such as promoting DNA repair, enhancing mitochondrial function, and reducing oxidative stress. It does so by deacetylating target proteins, which alters their activity.

   SIRT1 is also involved in regulating the response to caloric restriction, a well-established method of extending lifespan in many species. Caloric restriction increases NAD+ levels, which activates SIRT1 and other sirtuins, leading to enhanced cellular stress resistance and improved longevity.

2. **Other Longevity Genes**:

   Beyond sirtuins, several other genes have been identified that influence aging and longevity. These include genes involved in **autophagy**, a process by which cells degrade and recycle damaged components, and those involved in **insulin/insulin-like growth factor (IGF) signaling**, which regulates growth, metabolism, and aging. The **FOXO** family of transcription factors and **AMPK** (AMP-activated protein kinase) are also crucial regulators of aging and lifespan.

   Interventions that modulate these pathways, such as activating sirtuins or promoting autophagy, are being explored as potential strategies to extend lifespan and improve health during aging.

Aging is closely associated with a variety of chronic diseases that become more prevalent with age. These include:

1. **Neurodegenerative Diseases**:
   Alzheimer's disease, Parkinson's disease, and other neurodegenerative disorders are characterized by the progressive loss of neurons, often due to oxidative stress, mitochondrial dysfunction, and protein aggregation. For example, the accumulation of the protein **amyloid-beta** in Alzheimer's disease leads to oxidative damage and inflammation, contributing to cognitive decline.

2. **Cardiovascular Diseases**:
   Aging is a major risk factor for cardiovascular diseases, including **atherosclerosis** and **heart failure**. The accumulation of oxidative stress, mitochondrial dysfunction, and changes in blood vessel elasticity contribute to the decline in cardiovascular health with age.

3. **Osteoporosis**:
   Age-related bone loss, known as osteoporosis, occurs due to an imbalance in bone remodeling. The biochemical processes that regulate bone resorption and formation become dysregulated over time, leading to weakened bones and an increased risk of fractures.

4. **Cancer**:
   As discussed in Chapter 19, aging is associated with an increased risk of cancer. Accumulation of genetic mutations, telomere shortening, and cellular senescence contribute to the malignant transformation of cells.

Interventions and the Biochemistry of Anti-Aging Strategies

As our understanding of the biochemistry of aging deepens, several interventions have emerged with the potential to slow down or even reverse aspects of aging. Some of these include:

1. **Caloric Restriction**:

   Caloric restriction has been shown to extend lifespan in many species, including yeast, worms, and mice. It is believed to work by reducing oxidative stress, enhancing mitochondrial function, and activating longevity pathways such as SIRT1 and autophagy.

2. **Exercise and Physical Activity**:

   Regular exercise has numerous biochemical benefits, including improved mitochondrial function, reduced oxidative stress, and enhanced DNA repair. Exercise also helps maintain muscle mass and bone density, counteracting the effects of aging.

3. **Pharmacological Interventions**:

   Several drugs and compounds are being explored for their anti-aging potential. For example, **resveratrol**, a polyphenol found in grapes, has been shown to activate sirtuins and extend lifespan in model organisms. Other compounds, such as **metformin** and **rapamycin**, are being investigated for their potential to delay aging and age-related diseases.

4. **Gene Therapy and Stem Cell Therapy**:

   Gene therapy to activate longevity genes, or stem cell therapy to replace damaged tissues, holds promise as cutting-edge strategies for combating aging and age-related diseases.

Conclusion

Aging is a natural part of life, but the biochemical processes that drive aging are becoming increasingly understood. From oxidative stress and telomere shortening to the role of sirtuins and other longevity genes, biochemistry provides valuable insights into how aging occurs at the molecular level. Advances in biotechnology and pharmacology may eventually offer ways to slow aging, improve healthspan, and even extend lifespan. Understanding the biochemistry of aging is crucial for developing interventions that can promote healthier, longer lives.

Chapter 21: Biochemistry of Exercise and Performance

Exercise is a powerful force that influences the body's biochemistry, affecting everything from energy production to muscle function. Understanding the biochemical processes that occur during exercise can enhance athletic performance, optimize recovery, and support overall health. This chapter explores the biochemical changes that occur during exercise, the various energy systems the body utilizes, and how exercise adaptations influence muscle metabolism and performance.

The Biochemical Changes During Exercise

Exercise induces a wide range of biochemical changes that enable the body to meet the increased demand for energy. When we exercise, muscles consume ATP (adenosine triphosphate) for contraction, and the body must continually replenish this energy source. The efficiency and speed at which ATP is replenished depends on the intensity and duration of the exercise.

There are three primary energy systems in the body that provide ATP during exercise: the ATP-CP system, the glycolytic system, and the oxidative system. These systems operate at different intensities and durations, and each relies on distinct biochemical pathways to generate ATP.

Energy Systems: ATP-CP, Glycolytic, and Oxidative

1. **ATP-CP System (Phosphagen System)**

   This system provides immediate energy by breaking down stored ATP and phosphocreatine (CP) in muscles. It is the primary energy source for very short, high-intensity activities like sprinting or lifting heavy weights. ATP and CP are quickly used up, typically within 10-15 seconds of maximal effort, and the system requires rest for replenishment.

2. **Glycolytic System (Anaerobic Glycolysis)**

   The glycolytic system comes into play during moderate to high-intensity exercise lasting between 30 seconds and 2 minutes. This system breaks down glucose or glycogen into pyruvate, producing ATP in the process. However, the production of ATP via glycolysis occurs without oxygen (anaerobic) and results in the accumulation of lactate (lactic acid), which can contribute to muscle fatigue. As the intensity increases, lactate production exceeds its clearance, leading to a buildup that causes discomfort and limits performance.

3. **Oxidative System (Aerobic Metabolism)**

   The oxidative system is the most efficient energy system and is used during prolonged, moderate-intensity exercise. It relies on the presence of oxygen to fully oxidize carbohydrates, fats, and sometimes proteins into carbon dioxide and water, producing large amounts of ATP. This system takes longer to kick in but is capable of sustaining energy production for extended periods such as during long-distance running, cycling, or swimming.

Muscle fibers are the primary tissues that utilize these energy systems. There are two main types of muscle fibers: **Type I (slow-twitch)** and **Type II (fast-twitch)** fibers. Type I fibers rely primarily on aerobic metabolism and are used in endurance activities. Type II fibers, on the other hand, are adapted for anaerobic, high-intensity activities and generate energy through the glycolytic system.

When engaging in exercise, the body recruits different types of muscle fibers depending on the intensity and duration of the activity. For example, sprinting recruits fast-twitch muscle fibers, which are efficient at generating quick bursts of energy but fatigue rapidly. On the other hand, marathon running predominantly uses slow-twitch fibers, which are more resistant to fatigue but less powerful.

Adaptations to Exercise and Training

Regular physical training induces biochemical adaptations that enhance the body's ability to produce and utilize energy efficiently. These adaptations occur at the level of individual muscle fibers, as well as within metabolic pathways.

1. **Increased Mitochondrial Density**

   Aerobic training, such as long-distance running or cycling, increases the number of mitochondria in muscle cells. Mitochondria are the powerhouse of the cell and are responsible for ATP production via oxidative phosphorylation. More mitochondria mean the muscle can produce ATP more efficiently, improving endurance and performance.

2. **Improved Glycogen Storage**

   Training increases the muscles' capacity to store glycogen, which is the stored form of glucose. Enhanced glycogen storage allows athletes to sustain high-intensity efforts for longer periods, delaying fatigue. In endurance sports, this is particularly important as glycogen is a key fuel source.

3. **Enhanced Lactate Clearance**

   With regular anaerobic training, the body becomes more efficient at clearing lactate from muscles and converting it into usable energy. This reduces the negative effects of lactate accumulation and delays the onset of fatigue, allowing athletes to perform at higher intensities for longer durations.

4. **Capillary Density and Oxygen Delivery**

   Both aerobic and anaerobic training increase capillary density in muscle tissues. More capillaries mean better oxygen and nutrient delivery to muscles, improving performance in both endurance and strength-based activities.

5. **Increased Enzyme Activity**

   Exercise training also increases the activity of various enzymes involved in energy production, such as those involved in glycolysis, the citric acid cycle, and oxidative phosphorylation. This enhances the speed and efficiency at which energy is produced during exercise.

Nutritional Biochemistry for Optimal Performance

Proper nutrition plays a critical role in supporting the biochemical processes required for exercise. The body needs an adequate supply of macronutrients (carbohydrates, proteins, and fats) and micronutrients (vitamins and minerals) to fuel exercise and repair muscle tissue afterward.

- **Carbohydrates** are the primary fuel source for exercise, especially during high-intensity activities. Ensuring adequate carbohydrate intake before and after exercise helps replenish glycogen stores and maintain energy levels.
- **Proteins** are vital for muscle repair and growth. After exercise, consuming protein helps in the recovery process by promoting muscle protein synthesis and minimizing muscle breakdown.
- **Fats** are the primary energy source during low-intensity and long-duration exercise. The body's ability to efficiently oxidize fat increases with training, providing sustained energy for endurance activities.
- **Micronutrients**, such as vitamins and minerals, support metabolic processes and help prevent injuries and illnesses. For example, magnesium and calcium play critical roles in muscle contraction and relaxation.

Conclusion

Exercise and physical performance are intricately linked to biochemistry. The body's ability to efficiently generate and utilize energy systems underpins athletic performance, and training can induce powerful biochemical adaptations to improve efficiency and delay fatigue. Understanding the biochemical processes that fuel exercise provides athletes with the knowledge to optimize their training, performance, and recovery. Additionally, the role of nutrition and metabolic pathways in fueling the body for exercise highlights the importance of a balanced approach to diet and training. By mastering these biochemical principles, individuals can enhance their physical performance and achieve their fitness goals more effectively.

Chapter 22: The Role of Biochemistry in Nutrition

Nutrition is fundamental to the proper functioning of the human body, providing the necessary biochemical components to sustain health, promote growth, and maintain energy levels. The nutrients we consume are metabolized through complex biochemical pathways to produce the energy, building blocks, and regulatory molecules required for life. This chapter explores the biochemical roles of macronutrients, micronutrients, and various diets, highlighting their importance in supporting metabolic pathways and overall health.

Macronutrients and Their Biochemical Roles in the Body

Macronutrients — carbohydrates, proteins, and fats — are required by the body in large quantities because they provide the energy and raw materials necessary for cellular function. Each macronutrient is processed differently in the body, undergoing distinct biochemical transformations to meet metabolic needs.

1. **Carbohydrates**

   Carbohydrates are the body's primary source of energy. They include simple sugars, such as glucose, and more complex polysaccharides, such as starch and glycogen. When consumed, carbohydrates are broken down into monosaccharides, primarily glucose, which enters the bloodstream and is utilized by cells for energy. Glucose is metabolized through **glycolysis**, the **citric acid cycle**, and **oxidative phosphorylation** to produce ATP. Excess glucose is stored as glycogen in the liver and muscles or converted to fat if in excess.

   In addition to providing energy, carbohydrates are involved in cell signaling and the synthesis of nucleic acids and other essential molecules.

2. **Proteins**

   Proteins are essential for the growth, repair, and maintenance of tissues. They are made up of amino acids, some of which are essential (must be obtained from the diet), and others that are non-essential (can be synthesized by the body). Proteins play a critical role in enzyme catalysis, cellular structure, immune function, and neurotransmission.

   When proteins are digested, amino acids are released and enter various biochemical pathways. Some amino acids are used directly in protein synthesis, while others undergo **catabolism** for energy production. Excess amino acids are deaminated (removal of the amino group), and the remaining carbon skeletons are converted into glucose (via gluconeogenesis) or fatty acids for storage.

3. **Fats**

Fats (lipids) serve as an energy reservoir and provide insulation and protection for vital organs. The primary dietary fats include triglycerides, which are composed of fatty acids and glycerol. During digestion, triglycerides are broken down into free fatty acids and glycerol. These fatty acids are then oxidized through **beta-oxidation** to produce ATP, especially during low to moderate-intensity exercise or fasting.

Fats also play a crucial role in forming cellular membranes, especially phospholipids, and in the synthesis of signaling molecules like eicosanoids. Additionally, fat-soluble vitamins (A, D, E, and K) require fats for absorption and transport within the body.

Vitamins and Minerals: Biochemical Functions and Deficiencies

Micronutrients, including vitamins and minerals, are required in much smaller amounts than macronutrients but are equally essential for normal body function. They act as cofactors in enzyme reactions, contribute to the maintenance of cellular structures, and support metabolic pathways that regulate energy production, immune function, and the nervous system.

Vitamins

- **Vitamin A** is essential for vision, immune function, and skin health. It is involved in the synthesis of retinol and other compounds necessary for cellular communication.
- **Vitamin C** acts as an antioxidant and is vital for collagen synthesis, wound healing, and immune function.
- **Vitamin D** regulates calcium and phosphorus metabolism, supporting bone health and immune function.
- **Vitamin E** is a potent antioxidant that protects cells from oxidative damage and supports skin and eye health.
- **Vitamin K** is essential for blood clotting and bone health, as it helps activate proteins involved in calcium metabolism.

Minerals

- **Calcium** is crucial for bone and teeth formation, muscle contraction, and nerve function. It also plays a role in blood clotting.
- **Iron** is a key component of hemoglobin, the protein that carries oxygen in red blood cells. Iron deficiency can lead to anemia and impaired oxygen transport.
- **Magnesium** is involved in hundreds of enzymatic reactions, particularly those related to energy production, DNA and protein synthesis, and muscle function.
- **Sodium** and **potassium** maintain cellular fluid balance, regulate blood pressure, and support nerve and muscle function.

Deficiencies in these micronutrients can lead to various metabolic disorders, including anemia, osteoporosis, and compromised immune function.

The Biochemical Effects of Diets

The food we consume not only provides essential nutrients but also influences metabolic pathways that regulate energy balance, fat storage, and overall health. Various diets and eating patterns have different biochemical impacts, with some promoting efficient metabolism and others leading to metabolic imbalances or deficiencies.

1. **Ketogenic Diet**

   The ketogenic diet is a high-fat, moderate-protein, low-carbohydrate diet that forces the body into a state of **ketosis**, where it burns fat for energy instead of glucose. During ketosis, the liver converts fatty acids into **ketone bodies**, which can be used as an alternative energy source for the brain and other tissues. While this diet may support weight loss and improve metabolic health in some individuals, it can also lead to nutrient deficiencies if not properly balanced.

2. **Intermittent Fasting**

   Intermittent fasting involves cycling between periods of eating and fasting. During fasting, the body shifts from using glucose to utilizing fatty acids and ketone bodies for energy. This change in metabolic pathways can promote fat loss and improve insulin sensitivity. However, it is important to ensure adequate nutrient intake during eating periods to prevent deficiencies.

3. **Mediterranean Diet**

   The Mediterranean diet emphasizes fruits, vegetables, whole grains, legumes, olive oil, and lean protein sources such as fish and poultry. Rich in monounsaturated fats, fiber, and antioxidants, this diet has been shown to support cardiovascular health, reduce inflammation, and lower the risk of chronic diseases. The biochemical effects of the Mediterranean diet are linked to improved lipid profiles, enhanced insulin sensitivity, and better overall metabolic health.

4. **Plant-Based Diets**

   Plant-based diets, which emphasize fruits, vegetables, legumes, and whole grains, are associated with lower levels of chronic diseases such as heart disease, diabetes, and certain cancers. These diets provide a variety of phytochemicals, antioxidants, and fiber, which can positively influence biochemical pathways involved in inflammation, oxidative stress, and cellular repair.

Nutrition and Metabolic Pathways

Dietary components are integrated into metabolic pathways that regulate energy production, tissue maintenance, and growth. For instance, after carbohydrate intake, glucose enters glycolysis and the citric acid cycle to produce ATP. Fats are broken down in the liver and muscles through beta-oxidation, producing acetyl-CoA that enters the citric acid cycle. Proteins are degraded into amino acids, which are used in protein synthesis or converted into intermediates for energy production.

The biochemical regulation of these pathways is influenced by hormones such as insulin, glucagon, cortisol, and thyroid hormones. For example, insulin facilitates the uptake of glucose into cells and promotes glycogen storage, while glucagon stimulates the release of glucose from the liver during fasting.

Personalized Nutrition and Biochemistry

Understanding an individual's unique biochemical makeup is crucial for optimizing nutrition. Personalized nutrition takes into account factors such as genetics, age, activity level, and metabolic health to tailor dietary recommendations. Genetic variations can influence how individuals metabolize nutrients, making personalized approaches more effective in managing conditions such as obesity, diabetes, and heart disease.

Advancements in nutrigenomics — the study of the relationship between nutrition and genes — are paving the way for more individualized dietary recommendations. By understanding how specific nutrients affect gene expression and metabolic pathways, personalized nutrition can optimize health outcomes and prevent chronic diseases.

Conclusion

Biochemistry and nutrition are intricately linked, with nutrients serving as the biochemical building blocks that support metabolism and health. The macronutrients provide energy, the micronutrients support vital biochemical processes, and diets shape metabolic pathways that regulate growth, repair, and overall function. Understanding the biochemical roles of nutrients is essential for making informed dietary choices that support health and performance. By incorporating personalized nutrition strategies, individuals can optimize their biochemical pathways, improving metabolic health and reducing the risk of chronic diseases.

Chapter 23: Biochemistry and Biotechnology

Biotechnology, the application of biological systems and organisms in technology, plays a pivotal role in modern biochemistry. The integration of biochemistry and biotechnology has revolutionized medicine, agriculture, and environmental management, creating new possibilities for addressing global challenges. This chapter explores the foundational principles of biochemistry that underlie biotechnology, along with the applications that have transformed multiple industries.

The Role of Biochemistry in Biotechnology Applications

Biochemistry is the bedrock of biotechnology. By understanding the molecular mechanisms that govern cellular processes, biochemists have been able to develop innovative technologies that manipulate biological systems for a wide range of applications. These applications include the production of biopharmaceuticals, genetic engineering, diagnostic tools, and bioremediation, among others.

Biotechnology harnesses the biochemical processes that occur in living organisms, such as enzymes, proteins, and nucleic acids, to produce useful products and services. The knowledge of enzymatic reactions, molecular biology, and metabolic pathways is critical for the development of biotechnology solutions. Advances in this field have been driven by the ability to manipulate genetic material, producing products in a controlled and scalable manner.

Recombinant DNA Technology and Gene Cloning

One of the most transformative innovations in biotechnology is recombinant DNA (rDNA) technology, which allows scientists to combine DNA from different sources to create new genetic sequences. This process has numerous applications in medicine, agriculture, and environmental science.

In gene cloning, specific genes are isolated and inserted into plasmids, which are small, circular DNA molecules. These plasmids are then introduced into host organisms, such as bacteria or yeast, where the cloned gene can be expressed to produce proteins or other biochemically active molecules. This process has led to the large-scale production of valuable substances such as insulin, growth hormones, and vaccines.

For example, human insulin, once harvested from animal pancreases, is now produced through recombinant DNA technology, providing a consistent, affordable, and non-animal source of insulin for diabetic patients.

Biochemical Processes in the Production of Biopharmaceuticals

The production of biopharmaceuticals is one of the most prominent areas of biotechnology, and it relies heavily on biochemistry. Biopharmaceuticals, also known as biologics, include proteins, nucleic acids, or cells that are used to treat diseases. These products are typically made through processes such as cell culture, fermentation, and protein purification.

One critical aspect of biopharmaceutical production is the design and use of expression systems—cells or organisms that are engineered to produce a desired protein. For example, mammalian cells are often used to produce therapeutic proteins, as they can perform post-translational modifications like glycosylation that are necessary for proper protein function.

Biochemical engineering also plays a significant role in optimizing production processes, ensuring high yields of proteins or biologic products, while minimizing costs and maintaining the quality and safety of the final product.

Biotechnology in Agriculture

Biotechnology has transformed agriculture by enhancing crop yield, pest resistance, and nutritional value. Genetic modification techniques, such as the introduction of foreign genes into crops, have enabled the development of genetically modified organisms (GMOs) that are more resistant to diseases, pests, and environmental stressors.

One well-known application is the creation of Bt crops, which are genetically engineered to express a protein from the *Bacillus thuringiensis* bacterium that acts as a pesticide. These crops are less reliant on chemical pesticides, leading to both economic and environmental benefits.

Biochemistry also plays a critical role in plant breeding and genetic modification. Understanding plant metabolism, nutrient uptake, and stress responses allows scientists to design crops that are more resilient to climate change, pests, and diseases. Additionally, biotechnological innovations in biofertilizers and biopesticides, which use naturally occurring microorganisms to promote plant growth or control pests, are helping make agriculture more sustainable.

Biotechnology in Environmental Biochemistry

Biotechnology also plays an essential role in environmental protection and sustainability. Bioremediation, the use of microorganisms or plants to clean up polluted environments, is a major area where biochemistry and biotechnology intersect. Through biochemical processes, these organisms can degrade toxic substances, such as oil spills, heavy metals, and pesticides, into less harmful compounds.

For example, certain bacteria have the ability to break down oil in contaminated water, a process known as biodegradation. In this case, the bacteria metabolize the hydrocarbons in the oil, converting them into simpler compounds like carbon dioxide and water, effectively cleaning the environment.

Another area of environmental biotechnology is waste management. Microorganisms are employed to decompose organic waste, turning it into compost or biogas, which can be used as a renewable energy source. By understanding the biochemical pathways of these microorganisms, scientists are able to optimize these processes for maximum efficiency and effectiveness.

The Future of Biotechnology and Synthetic Biology

The field of biotechnology continues to evolve rapidly, with new advancements reshaping industries and societies. One of the most exciting areas is synthetic biology, which involves designing and constructing new biological parts, devices, and systems that do not occur naturally. This interdisciplinary field combines biology, chemistry, engineering, and computer science to create novel organisms or biological systems with specific, programmed functions.

Synthetic biology has the potential to revolutionize medicine, energy, and environmental management. For example, synthetic organisms could be designed to produce biofuels or biodegradable plastics, reducing reliance on fossil fuels and reducing pollution. In medicine, synthetic biology could enable the creation of tailor-made cells or tissues for regenerative therapies, offering personalized solutions for patients.

Biotechnology and synthetic biology also intersect with the growing field of personalized medicine, where biochemistry plays a role in tailoring medical treatments based on an individual's genetic makeup. This approach holds the potential to provide more effective and targeted treatments, minimizing side effects and improving patient outcomes.

Conclusion

Biochemistry and biotechnology are inextricably linked, each pushing the other forward in the pursuit of new solutions to global challenges. From the production of life-saving biopharmaceuticals to sustainable agriculture and environmental protection, the application of biochemistry in biotechnology is transforming the world. With continued advancements in genetic engineering, synthetic biology, and environmental biotechnology, the future of this field holds the promise of even greater innovations, improving lives and fostering sustainability.

As we look ahead, the synergy between biochemistry and biotechnology will remain a key driver of progress, unlocking new opportunities for human health, ecological balance, and technological advancement. Understanding the biochemical foundations of these technologies will be essential for harnessing their full potential in the coming years.

Chapter 24: Biochemistry in Clinical Practice

Biochemistry lies at the heart of modern medicine, providing the molecular understanding necessary for diagnosing diseases, monitoring health, and developing treatments. In clinical practice, biochemistry allows clinicians to assess the underlying biochemical abnormalities of various conditions and implement therapeutic strategies accordingly. This chapter explores the vital role biochemistry plays in clinical diagnostics, disease management, and the future of personalized medicine.

The Role of Biochemistry in Diagnostics and Clinical Testing

Biochemical analysis is fundamental in diagnosing a wide range of medical conditions, from metabolic disorders to infectious diseases and cancer. In clinical settings, biochemical tests are routinely used to measure concentrations of various substances in bodily fluids like blood, urine, and cerebrospinal fluid. These substances—such as glucose, proteins, enzymes, lipids, and electrolytes—are biomarkers that can indicate the presence, severity, and progression of disease.

For example, blood glucose levels are a primary marker for diagnosing and monitoring diabetes. Elevated levels of specific enzymes, such as alanine transaminase (ALT) or aspartate transaminase (AST), can indicate liver damage or disease. Biochemical tests also provide critical information in managing conditions like heart disease, kidney failure, and thyroid disorders.

One of the most important applications of biochemistry in diagnostics is the use of biomarkers—measurable indicators of a specific biological state. These biomarkers can be molecules, genes, or even specific patterns of metabolic activity. With advances in molecular biology, high-throughput technologies such as PCR, mass spectrometry, and ELISA (enzyme-linked immunosorbent assay) are increasingly employed to identify biomarkers for diseases, including infectious diseases, autoimmune conditions, and cancers.

Biomarkers and Laboratory Medicine

Biomarkers play an indispensable role in laboratory medicine, helping to identify the presence of disease and gauge its progression. In oncology, for instance, tumor markers such as prostate-specific antigen (PSA) for prostate cancer or CA-125 for ovarian cancer are widely used to monitor the effectiveness of treatment and detect recurrences. Furthermore, the discovery of novel biomarkers continues to improve early detection, particularly in the case of complex diseases like cancer, Alzheimer's disease, and cardiovascular conditions.

Emerging technologies, such as next-generation sequencing (NGS) and proteomics, are revolutionizing the identification of new biomarkers. These advancements allow for a more comprehensive understanding of the molecular signatures of diseases, enabling more precise and earlier diagnoses. For example, liquid biopsy—using blood samples to detect DNA or RNA fragments from tumors—is a promising non-invasive technique for detecting cancer at an early stage.

The integration of bioinformatics also aids in the interpretation of complex biochemical data, allowing for the identification of new biomarkers and the development of predictive models that guide clinical decision-making.

Biochemical Tests for Disease Diagnosis and Prognosis

Clinical biochemistry encompasses a broad array of tests that help diagnose and predict the course of disease. These tests assess the levels and activity of molecules in the blood, urine, or tissues, providing insights into the function of organs, the presence of infections, or the development of metabolic disturbances.

1. **Enzyme Activity Tests**: Elevated levels of certain enzymes can indicate organ damage. For example, increased levels of creatine kinase (CK) and troponin in the blood are indicative of myocardial infarction (heart attack), while elevated alkaline phosphatase (ALP) levels can suggest liver or bone disorders.
2. **Lipid Profiles**: The measurement of cholesterol levels (LDL, HDL) and triglycerides is used to assess cardiovascular risk and monitor patients with heart disease or dyslipidemia.
3. **Electrolyte Imbalances**: Sodium, potassium, calcium, and chloride levels are closely monitored in patients with kidney disease, dehydration, and electrolyte disturbances.
4. **Blood Gas Analysis**: Blood gas tests are used to evaluate respiratory and metabolic function, including measuring pH, oxygen, and carbon dioxide levels to assess conditions such as acidosis, alkalosis, and respiratory failure.
5. **Genetic Testing**: With advancements in genomic medicine, testing for inherited genetic mutations, such as those associated with cystic fibrosis or sickle cell anemia, has become a vital part of clinical diagnostics. Whole genome sequencing also offers a comprehensive view of a patient's genetic predispositions and disease risks.

6. **Hormonal Assays**: Biochemical tests can assess hormone levels in disorders of the endocrine system, such as thyroid dysfunction, diabetes, and adrenal diseases. Hormone panels help in diagnosing conditions like hyperthyroidism or hypoglycemia.

The ability to identify biochemical abnormalities early enables clinicians to initiate targeted treatments promptly, thereby improving patient outcomes.

Personalized Medicine and Pharmacogenomics

Personalized medicine is an innovative approach to healthcare that tailors medical treatment to the individual characteristics of each patient, particularly their genetic profile. By understanding the biochemical pathways affected by genetic variations, clinicians can optimize drug selection and dosing, minimizing side effects and maximizing therapeutic efficacy.

Pharmacogenomics is a branch of personalized medicine that focuses on how a patient's genetic makeup influences their response to drugs. Genetic variations can affect the metabolism of medications, altering their effectiveness or increasing the risk of adverse reactions. For example, individuals with variations in the CYP450 enzyme gene may metabolize certain drugs more slowly or quickly than others, requiring adjustments to drug dosages.

This approach has been transformative in fields such as oncology, where treatments like targeted therapies are becoming more common. For example, drugs like imatinib (Gleevec) for chronic myelogenous leukemia are specifically designed to target genetic mutations in cancer cells. Understanding these mutations on a molecular level allows clinicians to select the most appropriate and effective treatment for each patient.

In addition, personalized medicine incorporates advances in biomarkers, allowing for earlier and more accurate diagnoses, targeted therapies, and better management of chronic diseases like cardiovascular disease, diabetes, and autoimmune disorders.

The Future of Biochemistry in Clinical Healthcare

The future of biochemistry in clinical practice holds immense promise, particularly in the realms of diagnostics, personalized medicine, and gene therapy. With the continued integration of cutting-edge technologies, clinicians will be able to access increasingly precise and comprehensive biochemical data, improving diagnostic accuracy and therapeutic outcomes.

1. **Artificial Intelligence and Biochemistry**: Artificial intelligence (AI) and machine learning are already beginning to transform clinical biochemistry by enabling faster data analysis, personalized diagnostics, and even predictive modeling of disease progression. AI algorithms can analyze biochemical data to identify subtle patterns that might be missed by human clinicians, improving decision-making and the speed of diagnosis.

2. **Gene Therapy**: Advances in gene editing technologies like CRISPR-Cas9 are opening new doors for treating genetic diseases by correcting defective genes at the molecular level. While still in its early stages, gene therapy holds the potential to cure previously untreatable conditions by addressing their biochemical origins.

3. **Non-Invasive Monitoring**: As non-invasive technologies advance, future clinical practices may rely more on simple, non-invasive biochemical tests. For instance, continuous glucose monitoring devices, wearable sensors, and saliva tests are already offering more accessible ways to monitor a patient's biochemical status in real time, enabling more proactive health management.

4. **Regenerative Medicine**: Biochemistry will continue to play a crucial role in regenerative medicine, which focuses on repairing or replacing damaged tissues and organs. By understanding and harnessing the biochemical signals involved in tissue regeneration, scientists are making strides toward developing therapies for conditions like spinal cord injuries, heart disease, and liver failure.

Conclusion

Biochemistry in clinical practice is at the forefront of transforming healthcare by enabling precise diagnostics, targeted therapies, and personalized medicine. The integration of advanced technologies, such as genetic testing, AI, and gene therapy, is revolutionizing how diseases are detected, treated, and managed. As we look to the future, biochemistry will continue to evolve, paving the way for new breakthroughs in clinical healthcare that will improve patient outcomes and enhance quality of life. By mastering the molecular mechanisms that govern health and disease, biochemistry remains essential to advancing medicine and shaping the future of healthcare.

Chapter 25: Future Directions in Biochemistry

Biochemistry is a rapidly evolving field, with new discoveries and technological advancements continuously reshaping our understanding of life at the molecular level. As we move deeper into the 21st century, biochemistry is becoming increasingly intertwined with cutting-edge innovations in artificial intelligence (AI), nanotechnology, and biotechnology. These advancements have the potential to not only revolutionize medicine, health, and technology but also redefine our relationship with biology itself. In this chapter, we will explore the future directions in biochemistry and examine how emerging technologies and interdisciplinary approaches are pushing the boundaries of what we can achieve in both basic and applied biochemistry.

Cutting-Edge Research and Advancements in Biochemistry

The future of biochemistry lies in a deeper understanding of molecular mechanisms, the development of novel tools for analysis, and the ability to manipulate biological systems with precision. Some of the most exciting areas of research include:

1. **Structural Biology and Cryo-EM**: Advances in cryo-electron microscopy (cryo-EM) have transformed our ability to visualize molecular structures at atomic resolution. This technology has revolutionized drug discovery, protein engineering, and the study of molecular interactions. As cryo-EM becomes more sophisticated, we will be able to map the structures of complex macromolecular assemblies, such as multi-subunit enzymes and large membrane-bound complexes, with unprecedented clarity. This will pave the way for the design of highly specific therapeutic agents and precision medicine.
2. **Synthetic Biology**: Synthetic biology seeks to redesign and engineer biological systems to perform novel functions that are not found in nature. Researchers are working to create synthetic organisms, biosynthetic pathways, and new molecular machines that can be programmed to produce valuable compounds or perform specific tasks. For example, engineered bacteria could be used to produce pharmaceuticals, biofuels, or other important chemicals. Additionally, synthetic biology is poised to revolutionize fields such as agriculture, where engineered crops could be designed to have enhanced resistance to disease or climate change.

3. **Single-Cell Analysis**: Traditional biochemistry typically involves studying bulk populations of cells, but single-cell analysis allows for the investigation of individual cells and their molecular profiles. This is crucial for understanding cellular heterogeneity in tissues, disease progression, and the molecular mechanisms underlying stem cell differentiation. Single-cell RNA sequencing, for example, can provide insights into gene expression patterns in different cell types within the same tissue, offering a more detailed picture of cellular function and behavior.

4. **Metabolomics**: The study of metabolites, small molecules that are the end products of cellular processes, is rapidly advancing. Metabolomics allows researchers to monitor cellular metabolic states in real time, providing a comprehensive understanding of how cells respond to environmental changes, diseases, or drug treatments. This area of research has great potential for biomarker discovery, early disease detection, and the development of targeted therapies for metabolic disorders.

The Role of Artificial Intelligence in Biochemistry

Artificial intelligence is playing an increasingly important role in biochemistry by enabling the analysis of vast amounts of data generated by high-throughput experiments, such as genomic sequencing, proteomics, and metabolomics. AI, particularly machine learning (ML) algorithms, is being used to:

1. **Predict Protein Structures**: With the recent breakthroughs in AI models, such as AlphaFold, predicting the three-dimensional structures of proteins based on their amino acid sequences has become a reality. This has profound implications for drug design and the understanding of protein function. AI-driven predictions could allow for the rapid identification of potential drug targets and the design of molecules that can bind to these targets with high specificity.

2. **Drug Discovery and Development**: AI is being used to analyze massive databases of chemical compounds, identifying those that may have therapeutic potential for specific diseases. By leveraging predictive modeling, AI can streamline the drug discovery process, reducing the time and cost of bringing new drugs to market. Additionally, AI algorithms can predict the likely toxicity and efficacy of new compounds, making drug development more efficient and targeted.

3. **Personalized Medicine**: AI can help tailor medical treatments to individual patients based on their unique genetic makeup, lifestyle, and environmental factors. By analyzing patient data, AI can identify patterns and make predictions about how different treatments will affect the patient's biochemical pathways. This is particularly important in areas like oncology, where cancer treatment is increasingly becoming personalized based on the genetic profile of the tumor.

4. **Systems Biology and Computational Modeling**: AI is playing a key role in modeling complex biological systems and predicting how changes at the molecular level affect cellular behavior and disease progression. Through computational models, AI can simulate cellular networks, metabolic pathways, and gene regulatory circuits, providing insights into disease mechanisms and potential therapeutic strategies.

The Impact of Nanotechnology and Biotechnology

Nanotechnology and biotechnology are poised to further revolutionize the field of biochemistry. These technologies enable researchers to manipulate biological molecules and cells with extraordinary precision, opening up new possibilities for medical treatments, diagnostics, and environmental solutions.

1. **Nanomedicine**: Nanotechnology has the potential to transform drug delivery, enabling highly targeted therapies with reduced side effects. Nanoparticles can be engineered to deliver drugs directly to diseased cells, such as cancer cells, without affecting healthy tissue. This targeted approach can increase the efficiency of treatments and reduce the need for invasive procedures. Additionally, nanoparticles can be used as diagnostic tools, providing real-time imaging of cellular processes and detecting diseases at earlier stages.
2. **Gene Editing and CRISPR**: The CRISPR-Cas9 gene editing technology has made it possible to make precise modifications to the DNA of living organisms. This has vast implications for treating genetic diseases, such as cystic fibrosis and sickle cell anemia, by directly correcting the mutations that cause these disorders. CRISPR is also being used in cancer research, where it may help identify new therapeutic targets or enable the development of more effective immunotherapies.
3. **Bioprinting**: 3D bioprinting is an emerging technology that uses living cells as bio-inks to print three-dimensional tissues and organs. Bioprinting has the potential to revolutionize regenerative medicine by enabling the creation of custom tissues for transplantation, as well as the development of models for drug testing. This technology could also help address the shortage of organ donors by creating functional organs from a patient's own cells.

4. **Biopharmaceuticals**: Biotechnology continues to play a key role in the development of biopharmaceuticals—therapeutic products derived from biological sources. Recombinant DNA technology allows for the production of proteins, antibodies, and vaccines with applications in treating a wide range of diseases, including cancer, autoimmune disorders, and infectious diseases. The development of monoclonal antibodies and gene therapies has already transformed treatment strategies in oncology and immunology.

The Intersection of Biochemistry, Ethics, and Society

As biochemistry continues to evolve, it will raise significant ethical and societal questions, particularly in areas like gene editing, biotechnology, and personalized medicine. Some of the key ethical issues that biochemists and clinicians will face include:

1. **Genetic Privacy and Data Security**: The increasing use of genomic data to guide medical treatments and research raises concerns about privacy and the potential misuse of personal genetic information. Strict protocols and regulations will be needed to protect individuals' genetic data from unauthorized access and exploitation.

2. **Equity in Healthcare**: As personalized medicine becomes more prevalent, there is the potential for disparities in access to advanced treatments. Ensuring that all patients, regardless of socioeconomic status, have equal access to the latest biotechnological advancements will be a critical challenge.

3. **Ethics of Gene Editing**: The ability to edit the human genome raises profound ethical questions, especially when it comes to germline editing, which could affect future generations. While gene editing holds the potential to cure genetic diseases, its use also requires careful consideration of unintended consequences, long-term effects, and the potential for misuse.

4. **Environmental and Ecological Impacts**: The development of genetically modified organisms (GMOs) and synthetic biology products must be carefully evaluated for their potential impact on ecosystems. Ensuring that these innovations are safe for the environment will require rigorous testing and regulation.

The Potential for Biochemistry to Shape Future Innovations in Medicine, Health, and Technology

Biochemistry has already had a profound impact on medicine, health, and technology, and its future potential is even more exciting. From the development of targeted therapies for cancer to the creation of personalized medicines that cater to the genetic makeup of individual patients, biochemistry will continue to drive innovation in healthcare. As we enter a new era of precision medicine, bioengineering, and molecular biology, the next few decades promise to bring revolutionary advances that will shape the way we treat disease, enhance health, and understand the very nature of life.

As biochemistry continues to progress, it will provide us with the tools to solve some of humanity's most pressing challenges—from curing genetic diseases to improving food security and addressing climate change. The future of biochemistry is bright, and its potential to improve the human condition is limitless.

www.ingramcontent.com/pod-product-compliance
Lightning Source LLC
Chambersburg PA
CBHW082247220526
45469CB00009B/2904